Introductory Agroforestry

Introductory Agroforestry

Alok Kumar Patra
Professor (Agronomy)
All India Co-ordinated Research Project on Integrated Farming Systems
Odisha University of Agriculture & Technology
Bhubaneswar-751 003, Odisha

CRC Press
Taylor & Francis Group
Boca Raton London New York

CRC Press is an imprint of the
Taylor & Francis Group, an **informa** business

NEW INDIA PUBLISHING AGENCY
New Delhi-110 034

First published 2023
by CRC Press
4 Park Square, Milton Park, Abingdon, Oxon, OX14 4RN

and by CRC Press
6000 Broken Sound Parkway NW, Suite 300, Boca Raton, FL 33487-2742

© 2023 New India Publishing Agency

CRC Press is an imprint of Informa UK Limited

The right of Alok Kumar Patra to be identified as author of this work has been asserted in accordance with sections 77 and 78 of the Copyright, Designs and Patents Act 1988.

All rights reserved. No part of this book may be reprinted or reproduced or utilised in any form or by any electronic, mechanical, or other means, now known or hereafter invented, including photocopying and recording, or in any information storage or retrieval system, without permission in writing from the publishers.

For permission to photocopy or use material electronically from this work, access www.copyright.com or contact the Copyright Clearance Center, Inc. (CCC), 222 Rosewood Drive, Danvers, MA 01923, 978-750-8400. For works that are not available on CCC please contact mpkbookspermissions@tandf.co.uk

Trademark notice: Product or corporate names may be trademarks or registered trademarks, and are used only for identification and explanation without intent to infringe.

Print and electronic editions not for sale in South Asia (India, Sri Lanka, Nepal, Bangladesh, Pakistan, Afghanistan and Bhutan).

British Library Cataloguing-in-Publication Data
A catalogue record for this book is available from the British Library

ISBN: 9781032428789 (hbk)
ISBN: 9781032428819 (pbk)
ISBN: 9781003364726 (ebk)

DOI: 10.4324/9781003364726

Typeset in Times New Roman
by NIPA, Delhi

Preface

Agroforestry, as an age-old land use system has been in practice for thousands of years by farmers all over the world. But only in recent years, it has been developed as a science to help farmers increase the productivity, profitability and sustainability of production on their land by combining the best attributes of forestry and agriculture. Today, agroforestry has established itself as a viable approach of integrated land management system not only for meeting the deficits of food, fodder, firewood and timber but also for ecological considerations like soil conservation, biodiversity preservation, watershed protection, wasteland management, carbon sequestration and mitigation of climate change effects. In this backdrop, agroforestry has been recommended as a core subject in the curriculum of the state agricultural universities. Keeping this in view, an effort has been made to write a textbook on agroforestry.

This book is primarily based on the syllabus of 'Introductory Agroforestry' taught to under-graduate horticulture and forestry students. This book has been divided into ten chapters covering all aspects of agroforestry including concepts, definition, history, benefits and limitations, systems classifications, tree-crop interaction, planning and management, design & diagnosis, and propagation and management practices of multipurpose trees. Any suggestion to improve the contents of the book will be highly appreciated. I take full responsibility for any errors in this book. Any shortcomings may be intimated so that it will be taken care of.

In writing this book, the literature on agroforestry developed by various organizations and agencies like WAC, ICFRE, FAO, AFNETA, ICRISAT, CAFRI, CAZRI, IGFRI, etc. is freely used. I extend my sincere thanks to the authors and editors of various books, journals and periodicals which have been used as reference material in this book. Every care has been taken to cite the bibliographic references. However, any omissions, misrepresentations, incorrect citations or other mistakes that may have occurred are regretted.

I am grateful to my colleagues in the All India Coordinated Research Project on Integrated Farming Systems, Odisha University of Agriculture & Technology, Bhubaneswar for their help in various ways during the preparation of this textbook. I express my gratitude to the New India Publishing Agency, New Delhi for bringing out the book timely and nicely.

I am thankful to my wife Jharashree and daughter Prachurya for their support and encouragement. I also express my indebtedness and gratitude to my beloved parents who are a constant source of inspiration to me throughout my academic journey.

ALOK KUMAR PATRA
OUAT, Bhubaneswar

Contents

Preface .... *v*
Glossary .... *xi*

1 Agroforestry - A Sustainable Land Use System 1
    Definitions of agroforestry 1
    Key traits of agroforestry practices 3
    Premises of agroforestry 4
    Objectives of agroforestry 5
    Need of agroforestry 6
    Potential of agroforestry in India 8
    Benefits from agroforestry 10
    Limitations of agroforestry 11

2 Social Forestry 13
    Characteristics of social forestry 13
    Objectives of social forestry 14
    Scope of social forestry 14
    Components of social forestry 16
    Roadside plantation 18
    Railway line plantation 19
    Canal banks plantation 20

3 Status of Indian Forests 22
    Classification of forest 22
    History of Indian forests 24
    Present status of Indian forests 26
    Forest types of India 27
    Role of forests in farming systems 32

4 Agroforestry System, Subsystem and Practice 36
    Agrisilvicultural systems 37
    Silvipastoral systems 51
    Agrisilvipastoral system 54
    Other agroforestry systems 56

5 Tree Crop Interaction in Agroforestry 59
Nature of interactions 60
Factors affecting interactions 61
Types of interactions 62
Management options to neutralize negative interactions 67

6 Selection of Tree Crop Species for Agroforestry 68
Benefits from multipurpose trees 68
Selection of appropriate species 70
Characteristics of MPTs suitable for agroforestry 70
Role of MPTs in agroforestry 71
Suitable MPTs for agroforestry under different climatic conditions 77
Suitable MPTs for agroforestry under different soil types 78

7 Agroforestry Planning and Management 80
Factors affecting the selection of tree species 80
Tree propagation methods 82
Nursery management 86
Tree management 91
Agricultural crop management 98
Management of fruit trees in agroforestry systems 100
Constraints in adoption of agroforestry technology 101

8 Agroforestry Diagnosis and Design 104
Key Features of D & D 104
Agroforestry systems research process 105
Micro D & D 108
Methodological considerations in D & D 111

9 Agroforestry Policy and Projects 114
History of agroforestry research 114
International institutes working in forest conservation and research 115
Agroforestry research initiatives in India 122
National Agroforestry Policy, 2014 124

10 Management Practices of Multipurpose Trees 128
Mangium (*Acacia mangium*) 128
Sissoo (*Dalbergia sissoo*) 131
Khaira (*Acacia catechu*) 133

Eucalyptus (*Eucalyptus* spp.)..135
Casuarina (*Casuarina equisetifolia*)..137
Teak (*Tectona grandis*)...140
Neem (*Azadirachta indica*)...144
Tamarind (*Tamarindus indica*)..147
Phalsa (*Grewia* spp.)..149
Mulberry (*Morus alba*)..151
Poplar (*Populus deltoides*)..153
Oak (*Quercus* spp.) 158
Bamboo (*Dendrocalamus strictus*)...160

Selected References..165

Web Links..167

Index 168

Glossary

Agrisilviaquaculture: A form of agroforestry consisting of tree (woody perennial), agricultural crop and freshwater aquatic animal components.

Agrisilviculture: A form of agroforestry consisting of tree and crop components.

Agrisilvipasture: A form of agroforestry consisting of tree, crop and pasture/animal components.

Agroforestry: A collective name for land use systems and technologies where woody perennials (trees, shrubs, palms, bamboos, etc.) are deliberately used on the same land management units as agricultural crops and/or animals, in some form of spatial arrangement or temporal sequence.

Alkali soil: A soil that contains sufficient alkali (sodium) to interfere with the growth of most crop plants (pH 8.5 or higher).

Allelopathy: Any direct or indirect harmful effect that one plant has on another or mutually on each other through the production of chemical compounds that escape into the environment.

Alley cropping: A farming system in which arable crops are grown in alleys formed by trees or shrubs, established mainly to hasten soil fertility restoration and enhance soil productivity.

Annual plant: A plant that grows for only one season (or year) before dying, in contrast to a perennial, which grows for more than one season.

Aquasilviculture: A form of agroforestry consisting of tree and aquatic animals.

Arid climate: Climate in regions that lack sufficient moisture for crop production without irrigation. In cool regions annual precipitation is usually less than 25 cm. It may be as high as 50 cm in tropical regions. Natural vegetation is desert shrubs.

Bench terrace: Conversion of a steep slope into a series of steps with near-horizontal hedges and near-vertical walls between hedges, using retaining walls (rock) or steep banks (soil) for intensive cropping.

Biomass: The weight of material produced by a living organism or collection of organisms. The term is usually applied to plants to include the entire plant, or it may be qualified to include only certain parts of the plant, *e.g.*, above-ground or leafy biomass. Biomass is expressed in terms of fresh weight or dry weight.

Breast height: Universally adopted standard height for measuring girth or diameter of standing trees. It is 1.37 m (4 feet 6 inches).

Browse: The buds, shoots, leaves and flowers of woody plants which are eaten by livestock or wild animals.

Bund: An artificial earthen embankment made across sloping agricultural land to cut short lengthy soil slopes and reduce water runoff and soil erosion. It demarcates plot boundary or used for other purposes.

Bush fallow: A fallow of usually 3–10 years in which the natural vegetation regenerates which is composed principally of shrubs and young trees.

Buttress: An outgrowth from the base of a tree connecting it with the roots, especially common in tropical rainforest species.

C_3 plants: Species with the photosynthetic pathway in which the first product of CO_2 fixation is the 3-carbon molecule, phosphoglyceric acid and comparatively photosynthetically less efficient than C_4 plants.

C_4 plants: Species with the photosynthetic pathway in which the first product of CO_2 fixation is the 4-carbon molecule, oxaloacetic acid and photosynthetically more efficient than C_3 plants.

CAM (Crassulacean Acid Metabolism) plants: Species with stomata that open primarily at night, and have organic acids, especially malic acid, as the primary CO_2 fixation products.

Carbon/nitrogen ratio: The ratio of the weight of organic carbon (C) to the weight of total nitrogen (N) in a soil or organic material.

Carbon-stock: Carbon stored in vegetation or soil.

Catchment area: An independent unit of treated or untreated land area contributing runoff water to a reservoir.

Clean (or clear) **bole:** The part of a bole or stem which is free from branches.

Clone: A group of plants originating from a single plant by vegetative propagation and therefore all having the same genetic make-up.

Community forestry: A form of social forestry, where tree planting is undertaken by community people on common lands for fuelwood, timber, leaf fodder and other products that results in the improvement of community life.

Contour: An imaginary line connecting points of equal elevation on the surface of the soil.

Contour hedgerow: Dense, narrow hedges of woody perennials, planted along the contour of a slope to prevent soil erosion.

Contour ridge: A series of parallel ridges on the contour of cultivated land which has grass or shrubs planted on them to control soil erosion and improve water management.

Coppicing: Cutting of certain tree species close to ground level to produce new shoots from the stump. Also occurs naturally in some species if the trees are damaged.

Cover crop: Crops grown primarily to cover the soil and to reduce the loss of moisture due to leaching and erosion by wind and water.

Crop rotation: The growing of different crops on a piece of land in a preplanned succession, usually to prevent soil exhaustion from a demanding crop or to control pests and diseases.

Cropping pattern: The yearly sequence and spatial arrangement of crops or of crops and fallow on a given area.

Cropping system: The cropping patterns taken up for a given piece of land, or order in which the crops are cultivated on a piece of land over a fixed period of time and their interaction with farm resources, other farm enterprises and available technology which determine their make-up.

Crown: The canopy or top of a single tree or other woody plant that carries its main branches and leaves at the top of a fairly clean stem.

Cut-and-carry: Fodder or other plant products which are harvested and carried to a different location to be used or consumed.

Deciduous plant: A plant that sheds all or most of its leaves every year during a certain season.

Deforestation: Disturbance, conversion or wasteful destruction of forest lands.

Denitrification: The biochemical reduction of nitrate or nitrite to gaseous nitrogen, either as molecular nitrogen or as an oxide of nitrogen.

Diagnosis and Design (D & D): A systematic and objective methodology developed by International Centre for Research in Agroforestry to initiate, monitor and evaluate agroforestry programmes.

Dormancy of seed: The inability of seed to germinate, even under favourable growing conditions.

Drought: The absence of precipitation for a period long enough to cause depletion of soil moisture and damage to plants.

Ecosystem: All the plants and animals in a given area and their physical environment, including the interactions among them.

Erosion: The wearing away of the land surface by running water, wind, ice or other geological agents, including such processes as gravitational creep.

Evaluation: A process of assessing and monitoring results for conformity with agreed objectives in order to improve ongoing activities and plan better for future ones.

Evaporation: Loss of moisture from surfaces other than plants.

Evapotranspiration: The combined loss of water from a given area and during a specified period of time, by evaporation from the soil surface and by transpiration from plants.

Evergreen: Plants which retain their leaves and remain green throughout the year.

Exotic: A plant or animal species which has been introduced outside its natural range.

Extension forestry: The practice of forestry in areas devoid of tree growth and other vegetation situated in places away from the conventional forest areas with the objective of increasing the area under tree growth.

Fallow: Period of one to many years during which a field is not cropped, to allow soil fertility to be restored.

Farm forestry: Commercial tree growing by farmers on their own land.

Farming system: An appropriate combination of farm enterprises *viz.,* crops, livestock, forestry, fishery, apiary, poultry, and the means available to the farmer to raise them for profitability. It interacts adequately with the environment without dislocating the ecological and socioeconomic balance.

Fixed-lift pruning: Complete pruning of all branches below a prescribed point on the stem of the tree. The lift is specified as height from the ground and the choice of height is based on the shape of development attained by the stand.

Fluting: Stem shows irregular involutions and swellings. It is considered a serious defect for timber production.

Fodder: Parts of plants which are eaten by domestic animals, these may include leaves, stems, fruit, pods, flowers or pollen.

Foliage: The mass of leaves of plants, usually of trees or bushes.

Forage: Vegetative material in a fresh, dried, or ensiled state which is fed to livestock (hay, pasture, silage).

Forest fallow: A longer fallow, usually of several years, in which the naturally regenerated vegetation is composed principally of larger or mature trees.

Forest garden: A land-use form on private lands outside the village in which planted trees and sometimes additional perennial crops occur.

Grafting: The practice of propagating plants by taking a small shoot from one and attaching it to another so that the cambium layers from both are in contact and the transferred shoot grows as part of the main plant.

Grassland: A plant community in which grasses predominate; trees are either altogether absent or present in relatively small number.

Green manure: Green plant material incorporated into the soil to improve its fertility.

Groundcover: Living or non-living material which covers the soil surface.

Gully: A deep, narrow channel cut into the soil by erosion.

Hardening off: Measures taken to prepare plants raised in a nursery for planting out in the field.

Hedgerow (or hedge): A closely planted line of shrubs or trees, often forming a boundary or fence.

Herbaceous: A plant that is not woody and does not persist above ground beyond one season.

Homegarden: A land use form on private lands surrounding individual houses with a definite fence, in which several tree species are cultivated together with annual and perennial crops, often with inclusion of livestock.

Indigenous: Native to a specific area; not introduced.

Infiltration: The downward movement of water into the soil.

Infiltration rate: A soil characteristic determining or describing the maximum rate at which water can enter the soil under specified conditions, including the presence of an excess of water.

Inoculation: The process of introducing pure or mixed cultures of microorganisms into natural or artificial culture media.

Intercropping: Growing two or more crops in the same field at the same time in a mixture.

Interface: The area where there is positive or negative interaction between two entities, such as between rows of trees and rows of crops.

Internal rate of return (IRR): The maximum rate of interest that a project can repay on loans while still recovering all investment and opportunity costs; or, the earning power of money invested in a particular venture.

Land equivalent ratio (LER): The relative land area under sole crop required to produce the same yield as obtained under a mixed or an intercropping system at the same management level.

Land-use system: The way in which land is used by a particular group of people within a specified area.

Layering: Method to stimulate growth of roots on shoots or twigs in order to propagate the layered plant true to type.

Leaching (of nutrients): The dissolving and washing away of nutrients down the soil profile by the action of rain water.

Ley farming: Rotation of arable crops requiring annual cultivation and artificial pastures occupying field for two years or longer.

Lignin: The complex organic constituent of woody fibers in plant tissue that, along with cellulose, cements the cells together and makes substances resistant to decomposition and after some modification may become part of the soil organic matter.

Litter: Layer of decomposing plant material (leaves, branches, etc.) covering the ground, especially under trees.

Lopping: Cutting one or more branches of a standing tree or shrub.

Microclimate: The specific local climatic conditions near the ground or area around plants up to 2 m height, resulting from the modifications of the general climatic conditions by local differences in relief, exposure and cover, etc.

Mineralization: The conversion of an element from an organic form to an inorganic state as a result of microbial decomposition.

Mixed farming: Cropping systems which involve the raising of crops, animals and/or trees.

Mixed intercropping: Growing two or more crops simultaneously with no distinct row arrangement.

Monoculture: The repetitive growing of the same sole crop on the same land.

Mother tree: Tree from which seeds or vegetative parts are collected to be propagated.

Mulch: A layer of loose material on the soil to reduce moisture loss, moderate soil temperature and inhibit weed growth.

Multipurpose trees: Trees yielding one or more products and offering environmental benefits as well.

Multistoried (or multistoreyed): Relating to a vertical arrangement of plants so that they form distinct layers, from the lower (usually herbaceous) layer to the uppermost tree canopy.

Net present value (NPV): An indicator of a project's long-term value as estimated at the time of implementation. It is calculated by summing all the annual net costs or benefits over the prescribed life span of a project, discounted at a pre-selected rate.

Nitrogen cycle: The sequence of chemical and biological changes undergone by nitrogen as it moves from the atmosphere into water, soil and living organisms, and upon death of these organisms (plants and animals) is recycled through a part or all of the entire process.

Nitrogen fixation: The biological conversion of elemental nitrogen (N_2) to organic combinations or to forms readily utilized in biological processes, by nitrogen-fixing microorganisms. When brought about by bacteria in the root nodules of leguminous plants, it is referred to as symbiotic; if by free living microorganisms acting independently, it is referred to as non-symbiotic or free fixation.

Nutrient pump: A deep tree root system, that takes up nutrients from deep soil layers and brings them to the surface in the tree and its litter fall.

Opportunity cost: The true sacrifice incurred by the choice of a given action.

Overstory (or overstorey): The highest layer of vegetation, often the tree canopy, which grows over lower shrub or plant layers.

Perennial plant: A plant that grows for more than one year, in contrast to an annual, which grows for only one year (or season) before dying.

Pollarding: Cutting back the crown of a tree in order to harvest wood and browse to produce regrowth beyond the reach of animals and/or to reduce the shade cast by the crown.

Polypot: Polythene bag used as a pot to raise young plants.

Prick out: Transplant seedlings from seed trays or seedbeds into nursery beds or pots.

Propagule: A part of a plant that can give rise to a new plant.

Protective plant: Plants grown to protect crops, soil or land from adverse environmental factors.

Provenance: Germplasm from a single place of origin. Germplasm from different provenances of the same species can differ in ways such as growth habit, biomass production or drought hardiness.

Pruning: Cutting back plant growth, including side branches or roots.

Recalcitrant seed: Seed that has to be sown fresh because it loses its viability rapidly; *e.g.*, that of many large seeded tropical fruits.

Recreation forestry: The practice of forestry with the objectives of raising flowering trees and shrubs mainly to serve as recreation forests for the urban and rural population.

Reforestation: Establishment of trees on ground that has been recently cleared of trees.

Regeneration: The renewal of a forest crop by natural or artificial means

Resource capture: The processes by which plants obtain light, water and nutrients.

Rhizosphere: Portion of the soil in the immediate vicinity of plant roots in which the abundance and composition of the microbial population are influenced by the presence of roots.

Root ball: The network of roots and the soil clinging to them when a plant is lifted from the soil or removed from a container.

Root sucker: A shoot arising from the root of a plant.

Rotation (in forestry): The length of time between establishment and harvesting of a tree.

Runoff: The portion of the precipitation on an area which is not absorbed by the soil but finds its way into the streams after meeting the persisting demands of evapotranspiration and other losses.

Selective pruning: Removal of some of the branches on the stem at various levels above the ground according to a particular prescription.

Semiarid: Term applied to regions or climates where moisture is more plentiful than in arid regions but still definitely limits the growth of most crop plants. Natural vegetation in uncultivated lands in these areas is short grasses, shrubs and small trees.

Sequential cropping: Growing two or more crops in sequence on the same field. The succeeding crop is planted after the preceding crop has been harvested.

Shelterbelts: A wind barrier of living trees and shrubs established and maintained for protection of crop fields.

Shifting cultivation: System of land use in which the site of cultivation is regularly changed, with older sites reverting to forest or bush fallow.

Shrub: A woody plant that remains less than 10 meters tall and produces shoots or stems from its base.

Silvipastoral system: A form of agroforestry systems consisting of the trees (woody perennial) and pasture/animal components.

Slash-and-burn system: A technique for land clearing to prepare for planting of annual food crops or trees.

Social forestry: The practice of using trees and/or tree planting specifically to pursue social objectives, usually betterment of the poor, through delivery of the benefits to the local people.

Soil conservation: A combination of all management and land-use methods that safeguard the soil against depletion or deterioration caused by nature and/or humans.

Soil organic matter: The organic fraction of the soil that includes plant and animal residues at various stages of decomposition, cells and tissues of soil organisms, and substances synthesized by the soil population.

Sole cropping: One crop variety grown alone in pure stand at normal density.

Species: A taxonomic category below genus.

Stress: Any factor that disturbs the normal functioning of an organism.

Suckers: A side shoot from the roots of a plant; a side growth arising from an auxiliary bud.

Sustainable land use: Land use that maintains productivity of the land while conserving or enhancing the resources on which future production depends.

Taungya: Method of raising forest crops in combination with an agricultural crop; the agricultural use to which the land is put does not generally continue throughout the rotation of the forest crop but is confined to that period which ends with the closing of the canopy of the forest crop.

Thinning: Intermediate cuttings that are primarily aimed at controlling the growth of stands by adjusting stand density.

Tip pruning: Pruning of a branch at a point other than at its junction with the stem.

Topography: The physical description of land; changes in elevation due to hills, valleys and other features.

Transpiration: The loss of moisture from plants in the form of water vapour.

Tree garden: Multistoreyed agroforestry systems in which a mixture of several fruit and other trees are cultivated, sometimes with inclusion of annual crops.

Variable-lift pruning: Complete pruning of all branches below a prescribed variable point on the stem of the tree. This point may be specified either as a proportion of the height or as a diameter limit.

Watershed: A geohydrological unit or all the land and water area bounded by a divide which contributes runoff to a common point. Watershed is considered to be synonymous with 'catchment basin' and 'drainage basin'.

Wildings: Wildings are planting materials collected from natural vegetation, *e.g.*, seedlings or cuttings.

Windbreak: A strip of trees or shrubs or crop plants serving to reduce the force of wind and provide a protective shelter against wind.

Woody: Plants which consist in part of wood; not herbaceous.

Zero-grazing: Livestock production systems in which the animals are fed in pens or other confined areas and are not permitted to graze.

Chapter 1

Agroforestry - A Sustainable Land Use System

Agroforestry is a new name for a set of old practices. The farmers have been practicing agroforestry since ancient times. The general concept of agroforestry is to integrate trees and agriculture so as to create a more diversified landscape, while providing the producers with new environmental and economic benefits. In other words, agroforestry is a method of farming that allows trees and shrubs to grow along with crops and/or livestock, therefore blending agriculture and forestry in the same production system. In fact, man's association with forest is much older than with agriculture. First man was a food gatherer and hunter in forests. Then he realized that the seeds of the fruits he collected germinated, grew into plants and bore the fruits again and thus man started to cultivate foods. Man's desire to live in a community created settled agriculture.

Since then, the pressure on the agricultural lands has increased manifolds due to the increasing population, expansion of urban area and the industrialization process. The environment has also been disturbed. Soil is losing its productivity and the biodiversity is threatened. Farming community is trying all means to increase the land productivity. Chemical fertilizers and pesticides are applied in higher proportion, causing environmental pollution hazards. Under all these circumstances agroforestry has shown that besides sustainable agriculture it can also help promote a better environment. This relatively young science known as agroforestry was brought from the realm of indigenous knowledge into the forefront of agricultural research four decades ago and was promoted widely as a sustainability-enhancing practice that combines the best attributes of forestry and agriculture.

DEFINITIONS OF AGROFORESTRY

A strictly scientific definition of agroforestry should stress two characteristics common to all forms of agroforestry and separate them from the other forms of land use.

1. The deliberate growing of woody perennials on the same unit of land along with agricultural crops and/or animals, either in some form of spatial mixture or in some temporal sequence.
2. There must be a significant interaction (positive and/or negative) between the woody and non-woody components of the system, either ecological and/or economical.

Several definitions of agroforestry have been suggested since its conceptualization as a land use approach in 1977. But in short, agroforestry is generally perceived as the deliberate cultivation of several productive components, of which at least one is a woody perennial, on the same piece of land, combined either spatially or sequentially.

In the early 1980s International Council for Research in Agroforestry (ICRAF) defined agroforestry as "a collective name for land-use systems and technologies where woody perennials (trees, shrubs, palms, bamboos, etc.) are deliberately used on the same land management units as agricultural crops and/or animals, in some form of spatial arrangement or temporal sequence". This definition gained wider acceptance and it was used till the mid 1990s. Then in 1996 ICRAF redefined agroforestry as "a dynamic, ecologically based natural resource management system that, through the integration of trees on farms and in the agricultural landscape, diversifies and sustains production for increased social, economic and environmental benefits for land-users at all levels". Thus, in agroforestry systems there are both ecological and economical interactions between the different components (Lundgren and Raintree, 1982).

A few other definitions used in agroforestry literature are:

1. Agroforestry is a sustainable management system for land that increases overall production, combines agricultural crops, tree crops and forest plants and/or animals simultaneously or sequentially and applies management practices that are compatible with cultural patterns of local population.
2. Agroforestry is a land-use that involves deliberate retention, introduction, or mixture of trees or other woody perennials in crop/animal production field to benefit from the resultant ecological and economical interactions.
3. Agroforestry is a dynamic, ecologically based, natural resource management practice that, through the integration of trees on farms and in the agricultural landscape, diversifies and sustains production for increased social, economic and environmental benefits.

These definitions imply that

1. Agroforestry systems normally involve two or more species of plants (or plants and animals), at least one of which is a woody perennial.

2. An agroforestry system always has two or more outputs.
3. A cycle of an agroforestry system always requires more than one year for completion.
4. Even the simplest agroforestry system is more complex, ecologically (structurally and functionally) and economically, than monocropping system.

Additionally, there are three attributes which, theoretically, all agroforestry systems possess (Nair, 2008).

1. **Productivity:** Agroforestry systems aim to maintain or increase production of preferred commodities as well as productivity of the land.
2. **Sustainability:** Agroforestry can achieve and indefinitely maintain conservation and fertility goals by conserving the production potential of the resource base, mainly through the beneficial effects of woody perennials on soils.
3. **Adoptability:** Improved or new agroforestry technologies that are introduced into new areas should fit the social as well as environmental characteristics of the land use system for which it is designed and should also confirm to local farming system.

KEY TRAITS OF AGROFORESTRY PRACTICES

Agroforestry practices are intentional combinations of trees with crops and/or livestock which involve intensive management of the interactions between the components as an integrated agroecosystem. The four key characteristics - intentional, intensive, interactive and integrated - are the essence of agroforestry which distinguish it from other agricultural or forestry practices.

Thus, a land use practice to be called as agroforestry must satisfy all of these four criteria.

Intentional: Combinations of trees, crops and/or animals are intentionally designed and managed as whole units, rather than as individual elements which may occur in close proximity but are controlled separately.

Intensive: Agroforestry practices are intensively managed to maintain their productive and protective functions, and often involve annual operations such as cultivation, fertilization and irrigation.

Interactive: Agroforestry management seeks to actively manipulate the biological and physical interactions among the tree, crop and animal components. The goal is to enhance the production of more than one harvestable component at a time, while also providing conservation benefits such as control of soil erosion or protection of wildlife habitat.

Integrated: The tree, crop and/or animal components are structurally and functionally combined into a single, integrated management unit. Integration may be horizontal or vertical, and above- or below- ground. Such integration utilizes more of the productive capacity of the land and helps to balance economic production with resource conservation.

PREMISES OF AGROFORESTRY

The premises on which the concept of agroforestry is based are partly biological and partly socioeconomic.

Biological premises

Agroforestry has a beneficial effect on the soil through efficient nutrient cycling. The roots of trees take up nutrients from the soil, convert and utilize them for the production of plant material and then return them to the soil in the form of tree litter. This litter is transformed into humus and later incorporated into the soil. In a well managed agroforestry system, the relatively more efficient nutrient cycle minimizes the leakages of nutrients from the system.

Trees are generally deep-rooted than agricultural crops and are often able to trap and utilize nutrients that have been leached from the upper layers of the soil. Some tree species have the capacity of 'pumping' nutrients from layers that are not normally tapped by agricultural crops. The compacting effects of falling rain on the soil are reduced in an agroforestry system decreasing soil erosion and thus another possible source of leakage of nutrients from the system is plugged.

Agroforestry is a system of land management in which tree crops are grown together with agricultural crops, one of its objectives being to optimize and sustain the total yields of the component crops. Competition among the different components of the system is not great enough to affect the total productivity of the system in an adverse manner. Water, nutrients and light are the limiting factors in an agroforestry system. The forest and agricultural species that are utilized in the system should be compatible and should complement each other during most stages of their lives. More specifically, with respect to water they should be unequal in competitive ability; with respect to nutrient, they should vary in ability to utilize the nutrients in different forms; and, with respect to light, those species should be selected which display growth patterns, rates of growth, phenology, and architecture permitting maximum interception of light by both the agricultural and forest crops at any one time.

Socioeconomic premises

Forests are being felled in by farmers who require the land to produce food for their very existence. These areas are basically unsuitable for arable agriculture, either because of the inherent infertility of the soils, or because the sites are prone to accelerated erosion, or because of a combination of these two factors. The people who clear the forests to produce food are often not unaware of the possible deleterious effects of their practices upon the ecosystem in terms of erosion hazards, the occurrence of droughts and floods, and the possibility of soil fertility decline. Yet, despite their knowledge of these adverse consequences, they persist with such practice due to lack of suitable alternatives for their survival.

The failure to develop the marginal lands often leads to retardation in the rate of improvement of the general economy. The developmental and technological options are fewer in marginal areas than in most other ecosystems. When the biological influences and services of forests are considered along with the specific socioeconomic problems of those who exist in marginal areas, the technological package should include agroforestry systems. If the economic returns from the agroforestry systems are significant and if these are designed to optimize the joint productivity of wood and food from the same unit of land with careful choice of agricultural and forest species and suitable management practices, the socioeconomic developmental problems of the area would be addressed adequately.

OBJECTIVES OF AGROFORESTRY

The major objectives of agroforestry include:

- To reduce pressure on natural forests
- To manage land efficiently so that its productivity is increased
- To encourage tree plantation with agricultural crops and livestock to improve overall productivity, income and livelihoods of rural households, especially the marginal and small farmers
- To meet the raw material requirements of wood based industries and other small cottage industries in rural areas
- To raise the production of small timber used by villagers for agricultural implements, house constructions and other domestic purposes
- To generate employment opportunities for rural people throughout the year
- To protect and stabilize ecosystems, and promote resilient cropping and farming systems to minimize the risk during extreme climatic events
- To use available resources efficiently and economically

- To raise the production of food crops, legumes and tuber crops to meet the rapidly growing food requirements of people
- To raise the fodder production to meet the requirements of domestic animals
- To promote the production of vegetables, fruits, meat, egg and fish for nutritional security of resource-poor farmer families

NEED OF AGROFORESTRY

Agroforestry, in true sense, has been realized as a need of the day. It does not confine to the regional, geographical or agroclimatic boundaries. Agroforestry concept has got a universal application. Though several factors may contribute to the agroforestry interventions throughout the world, these factors are basically interdependent.

Decreasing land resources

There is an increased demand for food and fodder due to increase in population of human and cattle. However, the land resources are decreasing due to various reasons like conversion of agricultural lands to wastelands as a result of erosion, salinization and water logging; and infrastructural developments. There is hardly any scope to increase food production by increasing the area under cultivation. A management system therefore, needs to be devised that is capable of producing food from marginal agricultural land and is also capable of maintaining and improving the environment. The shrinking of per capita land availability, huge demand supply gap of various kinds of woods, food products as well as fodders are making agroforestry viable and an alternative land use option.

Limiting carrying capacity of the land

The carrying capacity of the arid and semiarid regions is overstressed. The consequence is destruction of environment leading to desertification. Agroforestry interventions hold the key to check soil erosion and leaching loss of nutrients and to improve the soil productivity through biological nitrogen fixation, organic matter addition and efficient nutrient cycling.

Overgrazing

The main reasons of overgrazing are increase in livestock population, increase in size of the herd and decrease in pasture availability. This problem is acute in arid and semiarid regions. Integrating cultivation of fodder tree species with suitable grasses in the wastelands would address the problem of overgrazing and thus check the desertification effectively.

Soil erosion and pollution

Soil is the most precious natural non-renewable resource. Soil erosion is the major cause of land degradation and loss of productivity. Erosion control has always started with tree and grass planting. Tree roots bind the soil and their leaves break the force of wind and rain on soil. Trees fight soil erosion, conserve rainwater and reduce water runoff. Trees also absorb dangerous chemicals and other pollutants that have entered the soil. Trees can either store harmful pollutants or change the pollutant into less harmful forms. Trees filter sewage and farm chemicals and reduce the effects of animal wastes. Thus, agroforestry practices are most suited for sustainability of soil productivity.

Overexploitation of land resource

Heavy fertilization coupled with high irrigation frequencies leads to soil loss, nutrient loss and degradation of land whereas under forest cover land upgradation is a continuous process. It restores soil and conserves moisture and thus, there is a gain from all angles. Taking advantages from both forest and agriculture, agroforestry concept itself becomes a profitable enterprise.

Fuelwood crisis

There is a global crisis of energy and man is striving hard to find out some alternative source of energy. Fuelwood is one of the established sources to meet energy requirement. About 90% people in the developing countries depend upon wood as source of fuel. But in these regions deforestation is five times more than afforestation. If this continues a time may come when there will be sufficient food but not the fuel to cook. So the only solution is to promote tree plantation through agroforestry.

Depletion of forest

Forest area is decreasing alarmingly due to demographic pressure and infrastructural developments causing thereby environmental pollution, ecological imbalance, global warming and climate change. Afforestation is costly and involves a long gestation period. There is also scarcity of land for creating new forests. Thus, an individual farmer cannot develop a new forest. But if both agriculture and forest are integrated (*i.e.*, agroforestry) then farmers can very easily adopt it as there will be no substantial reduction in agricultural output. Besides, all the utilities of forest will be available.

POTENTIAL OF AGROFORESTRY IN INDIA

Forest coverage

Forest cover in the country is 7,08,273 sq km, constituting 21.54% of its total geographical area against the ideal coverage of 33.33%. Out of this, very dense forest (>70% canopy density) constitutes 98,158 sq km (2.99%), moderately dense forest (40–70% canopy density) 3,08,318 sq km (9.38%) and open forest (<40% canopy density) constitutes 3,01,797 sq km (9.18%). The forest cover in the hilly districts is only 35.85% compared with the desired 66.66% area. The per capita availability of forests in India is one of the lowest, 0.064 ha against the world average of 0.64 ha and 1.07 ha for developed countries. The productivity of our forests is only 1.34 m^3/ha/year against the world average of 2.1 m^3/ha/year. Thus to bridge the gap between desired and available forest coverage in the country, agroforestry is the best intervention.

Utilization of wastelands and degraded lands

Areas presently not available for arable cropping can be put to agroforestry practices. According to the estimation of National Wasteland Development Board, 123 million hectare area of land is lying as wasteland in India. Out of this area water erosion affects the largest proportion followed by salinization and terrain deformation. The extent of degraded forests in the country is more than 40 million ha. Besides, about 50 million ha area is degraded due to mining activity. These areas can be reclaimed by adoption of suitable agroforestry practices. Large area is available in the form of farm boundaries and field bunds, where also agroforestry systems can be adopted.

Reduction in pressure on forest

Since land holding is becoming smaller and smaller due to demographic pressure, forest area in the vicinity of the thickly populated villages is diminishing with increasing human demands for fuel, fodder, small timber and other minor products met from the forest. Thus, by adopting agroforestry in the community lands near the villages, the pressure on natural forest could be greatly reduced.

Diversified products

Agroforestry provides diversified products such as fuel, fodder, fruits, fibre, timber, *etc*. Productivity is increased per unit area and per unit time. Agroforestry aims to maximize production of biomass of trees and agricultural crops. Tree and agriculture crop production system is more productive and is capable of meeting almost all the demands of timber, fodder, fruits, fibre and firewood.

Continuous revenue generation

The agroforestry plot remains usually productive for the farmer and generates continuous revenue, which is not feasible in arable land. Agroforestry also allows for the diversification of farm activities and makes better use of environmental resources.

Firewood production

About 87% of the annually harvested wood in India is used as firewood. In addition, at present in rural India 60–80 million tonnes of dry cow dung is utilized as fuel, equivalent to 300–400 million tonnes of freshly collected manure. Thus, there is a vast scope to meet the acute shortage of fuelwood through agroforestry.

Increase in quality grazing lands

The grazing lands in almost all parts of the country have to support animals beyond their carrying capacity. Repeated grazing by animals hardly leaves any vegetational element to survive unless specially protected. Inclusion of fodder tree species with suitable grasses in the agroforestry system will check overgrazing.

Employment opportunities

Agroforestry provides employment with relatively less investment and that too for unskilled rural community. It has a tremendous potential for rural employment generation due to great diversity of products from homegarden which provides opportunities for development of small scale rural industries and creation of off-farm employment and marketing opportunities.

Alternative land uses

Agroforestry provides strong incentives for adoption of conservation practices and alternative land uses and supports a collaborative watershed analysis approach to management of landscapes containing mixed ownerships, vegetation types and land uses. The only weapon that can be used in the war against hunger, inadequate shelter and environmental degradation is adoption of agroforestry practices.

Improvement in soil health

Trees and shrubs improve the physical properties of soils. Trees enhance water infiltration and water holding capacity of soils thereby reducing surface run-off and soil erosion. The repeated application of tree biomass increases the soil organic matter that leads to important increase in soil water retention capacity. The trees provide a favourable environment for soil microbes which in turn break down the biomass and release plant nutrients.

Insurance against climatic hazards

One of the most important contributions of agroforestry in general is to respond to climate change through sequestration of carbon in above ground plant biomass and below ground biomass in the soils. Trees and shrubs have the potential to reduce the impact of droughts, a common seasonal phenomenon in most of the developing countries like India where agriculture is mainly rainfed.

BENEFITS FROM AGROFORESTRY

Agroforestry is the system of developing agricultural land in combination with forestry technologies. Through this system, land with shrubs and trees are used to grow crop and livestock to encourage profitability, productivity, diversity and sustainability. There are numerous benefits of agroforestry as it encourages the adaptation of natural ecological processes within the commercial system. It helps farmers in terms of controlling land degradation, sheltering crop and livestock, improving their landscape and enhancing wildlife habitat. Benefits from agroforestry can be grouped under three broad categories - environmental, economic or social benefits.

Environmental benefits

1. Reduction of pressure on natural forests
2. More efficient recycling of nutrients by deep rooted trees on the site
3. Better protection of ecological systems
4. Reduction of surface runoff, nutrient leaching and soil erosion through impeding effect of tree roots and stems on these processes
5. Improvement of microclimate, such as lowering of surface soil temperature and reduction of evaporation of soil moisture through a combination of mulching and shading
6. Increment in soil nutrients through addition and decomposition of litter
7. Improvement of soil structure through the constant addition of humus from decomposed litter

Economic benefits

1. Increment in outputs of food, fuelwood, fodder, fertilizer and timber
2. Reduction in incidence of total crop failure, which is common to single cropping or monoculture systems

3. Increase in levels of farm income due to improved and sustained productivity

Social benefits

1. Improvement in rural living standards from sustained employment and higher income
2. Improvement in nutrition and health due to increased quality and diversity of food outputs
3. Stabilization and improvement of communities through elimination of the need to shift farm activities from one site to other

LIMITATIONS OF AGROFORESTRY

Agroforestry has different types of advantages but at the same time it also has some disadvantages.

1. Trees in agroforestry systems often compete with agricultural crops for light, water and nutrients from the soil which may reduce crop yields.
2. The use of farm machines is more difficult in the confined space in an agroforestry field.
3. This system is very difficult to manage and needs more accuracy with highly skillful management practices.
4. Increased susceptibility to pests and diseases often leads to dependence on potentially harmful pesticides.
5. Some of the ecological functions played by the trees in the natural forest may be lost when trees are used in an agroforestry system.
6. Damage to food crops during tree harvest operations.
7. Trees serve as hosts to insect pest and diseases that are harmful to agricultural crops.
8. Rapid regeneration by prolific trees which may displace food crops and take over the entire field.
9. Agroforestry systems are very labour intensive which may cause labour scarcity at times of other farm activities.
10. Longer period is required for tree components to mature and acquire an economic value.
11. Farmers are usually unwilling to displace food crops with trees, especially where land is scarce.
12. Agroforestry is more complex, less understood and more difficult to apply as compared to monocropping.

13. There is no government policy for tree felling in private lands. Farmers are often not allowed to harvest certain tree species grown by them in their own land which discourages strongly to go for tree cultivation.
14. Usually minimum support price for the tree products and other forest products is not fixed by any government agency. So the growers do not get good price for the produce from the tree components of agroforestry systems.
15. Although tree farming requires high initial investment and return is usually delayed there is no policy for financial support to the tree growers or agroforesters through nationalized banks.

Chapter 2

Social Forestry

The word 'social forestry' was coined by J. C. Westoby and used in the ninth Commonwealth Forestry Conference at New Delhi in 1968. In India the term was first used by the National Commission on Agriculture, Government of India in 1976, to denote tree raising programmes to supply firewood, small timber and minor forest products to rural population. Prasad (1985) defined social forestry as forestry outside the conventional forests which primarily aims at providing continuous flow of goods and services for the benefit of people. This implies that the production of forest goods for fulfilling the needs of the local people is social forestry. Thus, conceptually it deals with the people to produce goods such as fuel, fodder, small timber, etc. to meet the needs of local community particularly the underprivileged section (Shah, 1988).

CHARACTERISTICS OF SOCIAL FORESTRY

1. Social forestry is the forestry for the people and by the people. Here, people are the direct beneficiaries and it cannot develop without their full participation.
2. It is the forestry on a small scale. Social forestry is developed by an individual or a household or a group of households or a community. Here, the productivity inputs are limited and if the resources are made available either through Government or community then it can be increased. Social forestry activities are often divided into two classes, community forestry and farm forestry depending on the benefits of activity diverted towards community or part of the community or individuals.
3. The multipurpose tree species selected for plantation in the social forestry are fast growing, early maturing with both productive and protective benefits.
4. The social forestry may range from monocropping multiple use goals (*i.e.*, fast growing tree species for fuelwood, poles, small timber, fodder, etc.) to integrated cropping goals like agroforestry for getting wood, food, fodder and green manure, etc.

OBJECTIVES OF SOCIAL FORESTRY

1. To increase tree cover outside the forest areas
2. To encourage participation of people and institutions in plantation related activities
3. To increase the production of forest produce (small timber, firewood, fodder, fruits and other non-timber forest products) to meet the daily needs of the local population and thereby to reduce pressure on natural forests for supply of forest produce
4. To suitably use low productive and non productive lands through tree plantation
5. To rehabilitate the degraded forests so as to optimize their productivity and restore their potential to provide ecosystem goods and services on sustainable basis
6. To generate marketable surplus of forest products to yield cash incomes and improve the consumption level of village people
7. To produce raw materials for village level cottage industries
8. To increase crop yields through appropriate agroforestry models
9. To create livelihood opportunities to economically backward sections of rural communities by collection of dry leaves, fuelwood, dropped down seeds, fruits, etc.
10. To improve carbon stock in the tree cover outside forest areas
11. To increase employment opportunities for rural poor throughout the year by collection of seeds, preparation of nursery beds and raising seedlings, etc.
12. To increase the natural beauty of the landscape and to create recreational forests for the benefit of rural and urban population
13. To improve the environment for protecting agriculture from adverse climatic factors
14. To conserve and improve ecology and environment in the region

SCOPE OF SOCIAL FORESTRY

Scope of social forestry is very wide in India. Various kinds of lands which are available or can be made available for social forestry plantation include wastelands, roadsides, canal banks, ravine lands, marginal and sub marginal agricultural lands, degraded forest lands, urban unutilized areas, industrial complexes, backyards, etc.

Wastelands: At present, approximately 63.85 million hectares of land is lying as wastelands in India. Out of these lands, about 50% are non-forest lands, which can be made fertile again if treated properly. At least one third of such non-forest wastelands can conveniently be available for social forestry plantation.

Roadsides, canal banks and rail side lands: A large area of land is available for social forestry plantation along the roadside, canal bank and rail side.

Ravine lands: There are about 10 million hectares of ravine land, mainly confined to the states of Uttar Pradesh, Madhya Pradesh, Rajasthan and Gujarat. These lands can be managed through integrated approach of soil conservation and afforestation. The best use of ravine lands is social forestry.

Salt affected areas: Salt affected degraded lands can be reclaimed through social forestry plantation with suitable multipurpose tree species.

Marginal and sub marginal agricultural lands: These are the agricultural lands which are not physically suitable for growing of agricultural crops because of the low inherent fertility status of the soil. Mostly, these lands are owned by the economically poor people who are usually unable to use high input production system. Therefore, social forestry has a great potential in upgrading the fertility of such lands and uplifting the economic level of land owners.

Other agricultural lands: Arid and semiarid areas are frequently prone to moderate to severe droughts. Tree plantation or forestry in these areas has a great role in protecting agriculture, improving productivity, stabilizing production and ameliorating the environment.

Degraded forests: About 40 million hectares forest areas are degraded due to heavy pressure on them for fuelwood and overgrazing. These degraded forests can be revived through social forestry.

Urban areas: There is a scope for tree plantation in unutilized lands in urban areas such as compounds of public buildings, public parks, roadsides and even the private gardens.

Industrial complexes: In industrial complexes there is scope for tree planting. Normally, at the time of establishment of industries much more land is acquired than what is actually needed immediately. Thus, most of the industrial complexes have some land which is not put to any specific use. These unutilized lands can be put under social forestry plantation.

Backyard planting: Generally, in rural areas a majority of households grow trees in their backyards and still they have some unutilized space which can be brought under multipurpose tree plantation.

Temple groves: The temple compounds and premises of other places of worship can be utilized for tree plantation. Traditionally and culturally, temple groves are the best protected in our country.

School premises: The spare lands in school premises can be utilized for tree plantation. Creation of school forests has tremendous educational value for the students.

Graveyards: Graveyards or cremation grounds usually have some land which can be utilized for growing trees.

COMPONENTS OF SOCIAL FORESTRY

Social forestry can be categorized into five groups such as farm forestry, extension forestry, reforestation of degraded forests, community woodlots and recreation forestry. Besides roadside, railway line and canal banks plantations may also be included under social forestry.

Farm forestry

Farm forestry is the deliberate attempt of farmers towards planting or keeping trees in their farmlands in order to supplement daily farm/household needs, meet the contingencies and take market opportunities created by the dwindling supply of tree/forest products from the common lands. There are reports on trees being increasingly used as contingencies and also contributing to the household food security. Farm forestry was defined by National Commission on Agriculture (1976) as the practice of forestry in all its aspects in and around the farms or village lands integrated with other farm operations. The Commission noted that any programme of planting of trees in farms should serve the following objectives.

1. To supplement production of fuelwood and small timber to meet the increasing requirements of local people
2. To release cow dung for use as manure
3. To increase production of leaf fodder
4. To create a diverse ecosystem by having trees interspersed with cultivation

At present both commercial and non-commercial farm forestry is being promoted in one form or the other. Individual farmers are being encouraged to plant trees on their own farmland to meet the domestic needs of the family. In many areas the tradition of growing trees on the farmland already exists. Non-commercial farm forestry is the main thrust of most of the social forestry projects in the country today. It is not always necessary that the farmer grows trees for fuelwood, but very often they are interested in growing trees without any economic motive. They may want it to provide shade for the agricultural crops; as wind shelters; soil conservation or to utilize wasteland.

Extension forestry

Extension forestry is the practice of forestry in areas devoid of tree growth and other vegetation situated in places away from the conventional forest areas with the objective of increasing the area under tree growth. Thus, planting of trees on the sides of roads, canals and railways, along with planting on non-forest wastelands to increase the boundaries of forest is known as extension forestry. Under this project there has been creation of woodlots in the village common lands, government wastelands and panchayat lands. This can further be classified as

Mixed forestry: It is the practice of forestry for raising fodder grass with scattered fodder trees, fruit trees and fuelwood trees on suitable wastelands, panchayat lands and village common lands. Mixed forestry in the village wastelands and panchayat lands should be acceptable to people. Only quick growing and those products of immediate concern such as fodder and grass should be taken up with optimum input and technology.

Shelterbelts: Shelterbelt is defined as a belt of trees and/or shrubs maintained for the purpose of shelter from wind, sun, snow drift, etc.

Linear strip plantations: These are the plantations of fast growing tree species on linear strips of land on the sides of public roads, canals and railway lines. The National Commission on Agriculture recommended the plantations of fast growing tree species on the sides of roads, canals and railway lines. Various states have taken up the linear strip plantation recommended by the NCA. Such plantations have the capability to generate wood and other products whose utilization may help the local people to meet some of their basic requirements.

Rehabilitation of degraded forests

The degraded area under forests needs immediate attention for ecological restoration and for meeting the socioeconomic needs of the communities living in and around such areas. The NCA suggested reforestation of degraded forests to achieve the following objectives.

1. To grow short duration fuel and timber species for meeting the requirements of local people
2. To organize fuelwood supplies at reasonable rates which would prevent pilferage from neighbouring commercial forests

3. To tie up degraded forest areas with nearby rural and semi-urban centres for their requirements of fuelwood
4. To provide employment
5. To rehabilitate the degraded forests

Community woodlots

The community woodlots, consists of plantations of fuelwood species on community village lands, with intended objective of increasing a villager's access to fuelwood, fruits and fodder.

Recreation forestry

It is the practice of forestry with the objective of raising flowering trees and shrubs mainly to serve as recreation for the urban and rural population. This type of forestry is also known as **'aesthetic forestry'** which is defined as the practice of forestry with the objective of developing or maintaining a forest of high scenic value.

ROADSIDE PLANTATION

The main objectives of roadside or avenue plantation include provision of comfort to the travelers, aesthetics and landscape improvement, stabilization of roadsides improvement of ecological conditions and maximization of the productivity of the site to meet the requirements of the local people. Effective roadside plantations also check the movement of sand and dust to the road from adjoining areas and restrict soil erosion along stream banks or near the bridges. Each 30 m width of trees can absorb about 6 to 8 decibels of sound intensity. Thus to reduce the noise generated by high speed traffic on national highways to tolerable limits, about 20–30 m wide belts of trees and shrubs should be planted. Similarly, in cities 7–15 m wide belts of trees and shrubs may be required for this purpose. Evergreen trees are better for effective noise abatement.

The roadside plantation may be done by balanced line, unbalanced continuous or discontinuous line, sporadic or parking system. The balanced line system gives a continuous green wall of uniform size trees whereas unbalanced line systems give alternate avenues of different species interspersed by ornamental trees, either continuous or discontinuous. At some selected spots the parking system of planting may be adopted to develop picnic spots or resting places for travelers. Single row planting is normally done along village and district roads while more than one row can be planted in highway roads. In case of multiple rows planting, the first row should be of shade and/or ornamental trees and the remaining rows may be planted with fast growing trees species. The row spacing may be adjusted with respect to the position of telephone or electric lines or drains or other structures along the road. Trees should not be planted under telephone or electric lines. For shade trees having broad crown plant to plant spacing should be of 12 to 14 m while for other trees it may be 5 to 10 m. Planting at about half the desired spacing is usually practiced and when the crowns of trees begin to touch, less promising ones are thinned to leave the remaining trees at desired spacing. Too close spacing along hill roads and on the curves should be avoided. The trees should not obstruct the view at places where the pedestrians or domestic animals might usually be crossing the road near the villages.

The preferred species for roadside plantation include *Acacia arabica, Acacia auriculiformis, Ailanthus excelsa, Albizia lebbek, Albizia procera, Anthocephalus chinensis, Azadirachata indica, Bauhuinia purpurea, Bauhinia variegate, Butea monosperma, Cassia fistula, Cassia siamea, Dalbergia sissoo, Delonix regia, Erythrina* sp., *Ficus glomerata, Gliricidia sepium, Madhuca longifolia, Mangifera indica, Pongamia pinnata, Polyalthia longifolia, Prosopis juliflora, Samanea saman, Saraca indica, Syzygium cumini, Tamarindus indica, Terminalia arjuna, Terminalia catappa*, etc.

RAILWAY LINE PLANTATION

The main objectives of railway line plantation include stabilizing the railway track and protecting it against erosion, checking illegal encroachment of railway land, optimum utilization of the surplus land for tree cultivation to produce products needed by the local people and checking of the shifting sand in desert areas getting on to the railway track. The first row of trees is recommended to be planted at a distance of about 7.5 m from the centre of the track. The trees in the first row should be such that it may not attain a height more than their distance from the railway track so that in the event of wind through the tree tops do not reach the railway line and create traffic hazard. The inner side of the curves should not be planted to ensure clear visibility of the track. On either side of an unmanned level crossing a length of about 100 m should be left unplanted. No planting should be done under the telephone or electric lines. The spacing between the lines and between the trees in a line maybe 5 m, 3 m or 2 m depending upon the species to be planted, the fertility of the site and the objectives of the plantation. Commercial timber species of 30 years or more rotation may be planted at 5 m × 5 m spacing. Fast growing species for fuelwood, poles or fodder maybe planted at closer spacing of 2 m × 2 m or 2 m × 1 m. Ornamental trees should be planted near the towns, railway stations, road crossings and places of tourist interest. Trees with brittle stems or branches should not be planted especially in the first row. Wind firmness is a desirable character for the species for railway line plantation. The plantations along the railway lines are prone to accidental fires originating from engine spark especially in areas infested with tall grass or bush growth. The species to be planted in such areas should be able to resist such accidental fires. Hardy species like *Acacia catechu, Acacia arabica, Albizia procera, Azadirachata indica, Butea monosperma, Cassia fistula, Cassia javanica, Delonix regia, Ficus bengalensis, Madhuca indica, Mangifera indica, Prosopis juliflora, Saraca indica, Terminalia arjuna*, etc. are given preference for railway line plantation.

CANAL BANKS PLANTATION

The main objectives of canal bank plantation include stabilization of canal banks against erosion, checking of shifting of sand getting into the canal, provision of comfort to travelers using canal side road, utilization of the available land for tree cultivation and production of tree products, particularly fuelwood and small timber for local people, checking water logging in strips along canals and in adjoining areas and improvement of aesthetics in the area. The number of rows to be planted and the position of the first row depend upon the width and nature of the land available for planting. The spacing between the trees in a line depends upon the species selected and the desired end product from the plantation. The first row of trees should preferably be of shade trees. For planting in the first row, the trees with a strong taproot system should be preferred. The choice of species depends on climatic factors, objectives of planting, etc. The preferred species for canal banks plantation are *Acacia arabica, Cassia javanica, Dalbergia sissoo, Delonix regia, Eucalyptus* spp., *Ficus bengalensis, Jacaranda ovalifolia, Populus* spp., *Vateria indica*, etc.

DIFFERENCE BETWEEN SOCIAL FORESTRY AND AGROFORESTRY

S. No.	Social Forestry	Agroforestry
1.	Social forestry is a plantation made on lands outside conventional forest areas.	Agroforestry is a method of farming that allows trees and shrubs to grow along with crops and/or livestock, therefore blending agriculture and forestry in the same production system.
2.	It is the forestry of the people, by the people and for the people.	Agroforestry is a system which is rather localized in its concept for managing the unit of land for maximization of production of agricultural crop and forest trees complimentary with each other.

S. No.	Social Forestry	Agroforestry
3.	The major objectives of social forestry include supply of fuelwood to rural people, fodder for cattle of the rural population, protection of agriculture by creation of diverse ecosystems, arresting wind and water erosion, etc.	The major objectives of agroforestry is to reduce pressure on natural forests and to encourage tree plantation with agricultural crops and livestock to improve overall productivity, income and livelihoods of rural households, especially the marginal and small farmers.
4.	Social forestry can be categorized into five groups such as farm forestry, extension forestry, reforestation of degraded forests, community woodlots and recreation forestry. Besides roadside, railway line and canal banks plantations may also be included under social forestry.	There may be several agroforestry practices like agrisilviculture, silvipastoral, hortisilviculture, hortisilvipastoral, etc. depending upon the components included in a system.
5.	Planting of trees on massive scale is done on vacant land, community land, roadside, railway track and even degraded reserve forest.	Agroforestry is practiced mostly in farmer's field/own land.
6.	Trees and shrubs are to be used to harvest multiple products.	It involves integration of two or more than two components in a cultivated land.
7.	Social forestry is short rotation forestry. Return from a social forest is relatively late.	From an agroforestry system the returns are quick from agricultural crop components.

Chapter 3

Status of Indian Forests

A forest is a large area dominated by trees. Many definitions of forest are used throughout the world, incorporating factors such as tree density, tree height, land use, legal standing and ecological function. Although a forest is usually defined by the presence of trees, under many definitions an area completely lacking trees may still be considered a forest if it grew trees in the past, will grow trees in the future, or was legally designated as a forest regardless of vegetation type. According to the Food and Agriculture Organization, forests covered 4 billion hectares or approximately 30% of the world's land area in 2006. Forests are the dominant terrestrial ecosystem of earth, and are distributed around the globe. Forests account for 75% of the gross primary production of the earth's biosphere, and contain 80% of the earth's plant biomass

CLASSIFICATION OF FOREST

Forests have been classified in different ways. Forests can be classified on the basis of method of regeneration, age, composition, object of management, ownership, canopy density, stand density or growing stock.

Method of regeneration

Forests may be high forest or coppice forests. A high forest is regenerated from seed and a coppice forest is regenerated by vegetative means such as coppicing or root suckers. When the regeneration is obtained naturally, the forests are called natural forests. When it is obtained artificially, the forests are called man-made forests or plantations.

Age of trees

Forests may be described and considered from the standpoint of the age classes of which they are composed. Forests composed of even-aged woods and applied to a stand consisting of trees of approximately the same age are called even aged or regular forests. Differences up to 25% of the rotation age are usually allowed in cases where forest is not harvested for 100 years or more. Forests composed of trees of markedly different ages and applied to a stand in which individual stem vary widely in age are called uneven aged or irregular forests.

Composition

A forest may be a pure forest or mixed forest. A forest in which at least 50% of the trees in the main canopy are of single species is called pure forest. A mixed forest is composed of trees of two or more species intermingled in the same canopy and no species has more than 50% of the trees in the canopy.

Objectives of management

Forests can be classified according to their expected uses or objectives of management.

1. Forest managed primarily for its produce is termed as production forest. It is also sometimes referred to as national forest, i.e., a forest which is maintained and managed to meet the needs of the defence, communication, industry, and other general purposes of public importance.
2. Protection forests are those which are managed primarily for ameliorating climate, checking soil erosion and floods, conserving soil and water, regulating streamflow and increasing water yields and exerting other beneficial influences.
3. Farm forest is a forest raised on farms and its adjoining area either as individual scattered trees or a collection of trees to meet the requirement of fuel and fodder of the farmers and to have a beneficial influence on agriculture.
4. Forest raised on village wasteland to supply fuel, small timber, fodder, etc., to the village communities is called fuel forest.
5. Forest which is managed only to meet the recreational needs of the urban and rural population is termed as recreational forest.

Ownership and legal status

State forest is a forest owned by state. On the basis of legal status, state forests are further classified as reserved forest, protected forest, village forest, communal forest or panchayat forest.

1. Reserved forest is an area so constituted under the Indian Forest Act 1927 or other forest law.
2. Protected forest is an area subject to limited degree of protection under the provision of chapter IV of the Indian Forest act 1927.
3. State forest assigned to a village community under the provision of the Indian Forest Act 1927 is called village forest.

4. Communal forest is a forest owned and generally managed by a community such as a village, town, tribal authority or local government, the members of which share the produce.
5. Panchayat forest is forest where management is vested in a village panchayat.

Canopy density

The forest which has more than 70% canopy density is called very dense forest. Moderately dense forest has 40–70% canopy density and open forest has less than 10–40% canopy density. Scrub (degraded forest lands) has canopy density <10%.

Growing stock

A normal forest is an ideal forest with regard to growing stock, age class distribution and increment and from which the annual or periodic removal of produce equals to the increment and can be continued indefinitely without endangering future yields. Abnormal forest is one which is not normal, i.e., growing stock, age class, distribution of stems, increment, etc. are either in excess or more usually in deficit than the normal forest.

Stand density

The density of stocking expressed in number of trees, basal area, volume, or other criteria, on a per-hectare basis.

1. Forests in which all the growing space is effectively occupied but which still have ample room for development of the crop trees is called fully stocked forests.
2. Forests in which the growing space is so completely utilized that growth has slowed down and many trees, including dominants, are being suppressed are called overstocked forests.
3. Forests in which the growing space is not effectively occupied by crop trees are called understocked forests.

HISTORY OF INDIAN FORESTS

The epics Ramayana and Mahabharata give an attractive description of forests. Ancient Hindu culture is said to have evolved in the forest. However, the earliest indication of forestry administration in India is found in 300 BC. It was during the reign of Chandragupta Maurya. A Superintendent of Forests looked after forests and wildlife. Later Ashoka continued the process. Much importance was also given for planting trees along roadsides.

The Mughals were interested in trees for gardening. They also showed interest in avenue plantations. They therefore displayed an aesthetic and utilitarian approach to plants. But they lacked any comprehensive understanding of forests. They lacked definitive approach for their preservation, propagation, protection and improvement. During this period the forests were reclaimed for agriculture. Parts of the farmer community were pushed back into the forests due to the Mughal invasion. They took up shifting cultivation and this damaged the forests.

Heavy destruction of forests also occurred in the later part of the 18th and early part of the 19th century. Europeans carried away much of the produce. In the early years of British Raj, large indents were made on the timber wealth of the country. The teak forests along the coast of Malabar were over-exploited. The timber was supplied to meet the requirement of the British Navy. Over-exploitation followed appointment of a commission in 1800 to inquire into the availability of teak. Sandalwood trees of south India were exploited for their way to European markets.

In 1840, the British colonial administration promulgated an ordinance called Crown Land (Encroachment) Ordinance. This ordinance targeted forests in Britain's Asian colonies, and vested all forests, wastes, unoccupied and uncultivated lands to the Crown. The Imperial Forest Department was established in India in 1864 and the British state monopoly over Indian forests was first asserted through the Indian Forest Act of 1865. This law simply established the government's claims over forests.

The British colonial administration then enacted a further far-reaching Forest Act of 1878, thereby acquiring the sovereignty of all wastelands which in its definition included all forests. This Act also enabled the administration to demarcate reserved and protected forests. These colonial laws brought the forests under the centralised sovereignty of the state. Sir Dietrich Brandis, the Inspector General of Forests in India from 1864 to 1883, is regarded as the father not only of modern scientific forestry in India, but also as the 'father of tropical forestry'.

After independence in 1952, the Indian government nationalised the forests which were earlier with the *zamindars*. India also nationalised most of the forest wood industry and non-wood forest products industry. Over the years, many rules and regulations were introduced by India. In 1980, the Conservation Act was passed, which stipulated that the central permission is required to practice sustainable agroforestry in a forest area. These nationalisation wave and laws intended to limit deforestation, conserve biodiversity, and save wildlife. However, India's rural population and impoverished families continued to ignore the laws, and use the forests near them for sustenance. Thus, deforestation increased, biodiversity diminished and wildlife dwindled.

India launched its National Forest Policy in 1988. This led to a programme named Joint Forest Management, which proposed that specific villages in association with the forest department will manage specific forest blocks. In particular, the protection of the forests would be the responsibility of the people. Since 1991, India has reversed the deforestation trend. By 1992, seventeen states of India participated in Joint Forest Management, bringing about 2 million hectares of forests under protection.

India's 0.6% average annual rate of deforestation for agricultural and non-lumbering land uses in the decade beginning in 1981 was one of the lowest in the world. India's forest cover grew at 0.2% annually over 1990–2000, and has grown at the rate of 0.7% per year over 2000–2010, after decades where forest degradation was a matter of serious concern. From 1990–2000, India was the fifth largest gainer in forest coverage in the world; while from 2000–2010, it was the third largest gainer in forest coverage.

PRESENT STATUS OF INDIAN FORESTS

India is one of the ten most forest-rich countries of the world along with Russia, Brazil, Canada, United States of America, China, Democratic Republic of the Congo, Australia, Indonesia and Sudan. Together, India and these countries account for 67% of total forest area of the world. Forestry in India is more than just about wood and fuel. India has a thriving non-wood forest products industry, which produces latex, gums, resins, essential oils, flavours, fragrances and aroma chemicals, incense sticks, handicrafts, thatching materials and medicinal plants. About 60% of non-wood forest products are consumed locally. About 50% of the total revenue from the forestry industry in India is in non-wood forest products category.

According to the India State of Forest Report 2017, the total forest cover is 7,08,273 sq km, which is 21.54% of the total geographical area of the country. Forest and tree cover combined is 8,02,088 sq km, which 24.39% of the total geographical area. The increase in the forest cover has been observed as 6,778 sq km and that of tree cover as 1,243 sq km over the 2015 assessment. The increase in forest cover has been observed in very dense forest which absorbs maximum carbon dioxide from the atmosphere. It is followed by increase in forest cover in open forest. About 40% of country's forest cover is present in 9 large contiguous patches of the size of 10,000 sq km, or more.

India targets bringing 33% of its geographical area under forest cover. As many as 15 states and union territories have forest cover exceeding 33% of their geographical area. Out of these, seven states (Mizoram, Lakshadweep, Andaman & Nicobar Islands, Arunachal Pradesh, Nagaland, Meghalaya and Manipur) have more than 75% forest cover while eight states (Tripura, Goa, Sikkim, Kerala, Uttarakhand, Dadra & Nagar Haveli, Chhattisgarh and Assam) have forest cover between 33 and 75%. Top 5 states with maximum increase in forest cover are Andhra Pradesh (2141 sq km), Karnataka (1101 sq km), Kerala (1043 sq km), Odisha (885 sq kms) and Telangana (565 sq kms).

Canopy-wise forest cover

Class	Area (square km)	Percentage of geographical area
Very dense forest	98,158	2.99
Moderately dense forest	3,08,318	9.38
Open forest	3,01,797	9.18
Total forest cover*	**7,08,273**	**21.54**
Scrub	49,979	1.40
Non-forest	25,33,217	77.06
Total geographical area	**32,87,469**	**100.00**

*Includes 4,921 sq km under mangrove cover
Source: India State of Forest Report, 2017

Three states with maximum forest cover in terms of area are Madhya Pradesh (77,414 sq km) Arunachal Pradesh (66,964 sq km) and Chhattisgarh (55,547 sq km). Three states with highest forest cover in terms of percentage of geographical area are Lakshadweep (90.33%), Mizoram (86.27%) and Andaman & Nicobar Islands (81.73%).

However, in the northeast region, the 2017 assessment shows an actual decrease of forest cover. Out of the eight states of the region, only Assam and Manipur have registered an increase in forest cover. While Assam registered an increase of 567 square km, for Manipur it was 263 square km forest area over the 2015 assessment. Major five states where forest cover has decreased are Mizoram (531 sq km), Nagaland (450 sq km), Arunachal Pradesh (190 sq km), Tripura (164 sq km) and Meghalaya (116 sq km). The main reasons for decrease are shifting cultivation, rotational felling, other biotic pressures, diversion of forest lands for developmental activities, submergence of forest cover, agriculture expansion and natural disasters.

FOREST TYPES OF INDIA

India has a diverse range of forests, from the rainforest of Kerala in the south to the alpine pastures of Ladakh in the north, from the deserts of Rajasthan in the west to the evergreen forests in the north-east. Climate, soil type, topography, and elevation are the main factors that determine the type of forest. Forests are classified according to their nature and composition, the type of climate in which they thrive, and its relationship with the surrounding environment. There are 6 major groups, namely, moist tropical, dry tropical, montane sub-tropical, montane temperate, sub-alpine, and alpine, further subdivided into 16 major types of forests.

Moist tropical forests

These are further classified into 4 types on the basis of relative degree of wetness.

Tropical moist evergreen forests: These are also called tropical rain forests. In India such forests are found in very wet regions receiving more than 250 cm average annual rainfall. These are climatic forests having luxuriantly growing lofty trees which are more than 45 metres in height. The shrubs, woody climbers and epiphytes are abundant because of high rainfall. These forests are found in Andaman and Nicobar Islands, western coasts and parts of Karnataka, Assam and Bengal.

Tropical moist semi-evergreen forests: These forests are found along the western coasts, eastern Odisha and upper Assam where annual rainfall is between 200 and 250 cm. They are characterised by giant and luxuriantly growing intermixed deciduous and evergreen species of trees and shrubs. The important species include *Terminalia, Bambusa, Dipterocarpus, Garcinia, Sterculia, Mallotus, Albizzia, Shorea, Bauhinia*, etc. Orchids, ferns, some grasses and several other herbs are also common.

Tropical moist deciduous forests: These cover an extensive area of the country receiving annual rainfall of 100 to 200 cm, spread over most part of the year. The dry periods are of short duration. Many plants of such forests show leaf-fall in hot summer. The forests are found along the wet western side of the Deccan plateau, Andhra, Gangetic plains and in some Himalayan tracts extending from Punjab in west to Assam valley in the east. The forests of south India are dominated by *Tectona grandis, Terminalia paniculata, Grewia tilliaefolia, Dalbergia latifolia, Adina cordifolia*, etc. In north, they are dominated by *Shorea robusta, Terminalia tomentosa, Dellenia* species, *Eugenia* species, etc. These forests produce some of the most important timbers of India.

Littoral and swamp forests: Littoral and swampy forests include beach forests, tidal forests or mangrove forests, and fresh water swamp forests.

Beach forests: The beach forests are found all along the sea beaches and river deltas. The soil is sandy having large amount of lime and salts but poor in nitrogen and other mineral nutrients. Ground water is brackish, water table is only a few metres deep and annual rainfall varies from 75 cm to 500 cm.

The common plants of these forests are *Casuarina equisetifolia, Phoenix, Callophyllum littoralis, Thespesia, Pongamia, Cocos nucifera*, etc.

Tidal or mangrove forests: Tidal forests grow near the estuaries or the deltas of rivers, swampy margins of islands and along sea coasts. The soil is formed of silt, silt-loam or silt-clay and sand. The plants are typical halophytes which are characterised by presence of prop roots with well developed knees for support and pneumatophores and viviparous germination of seeds. Major species in these forests include *Rhizophora mucronata, Avicennia alba, Avicennia officinalis, Bruguiera parviflora, Kandelia candel, Xylocarpus granatuns, Xylocarpus molluccensis, Ceriops tagal, Ceriops decandra, Excoecaria agallocha, Sonneratia acida, Sonneratia caseolaris, Lumnitzera racemosa, Aegiceras carniculatum, Heritiera minor, Bruguiera conjugata, Cynometra ramiflora, Amoora cuculata, Phoenix paludosa*, etc. Tidal forests may be distinguished into four types with overlapping constituent species; tree mangrove forests, low mangrove forests, salt water forests and brackish water forests. Tree mangrove forests occur on both east and west sea coasts. The best development occurs in Sundarbans. The forest floor is flooded with salt water daily. Low mangrove forests grow on soft tidal mud near estuaries, which is flooded by salt water. Forest is dense but the trees with leathery leaves attain maximum height of 3-6 m. Salt water mangrove forests occur beyond tree mangrove forests in big river deltas where the ground is flooded with tidal water. Brackish water mangrove forests grow near the river deltas where forest floor is flooded with water at least for some times daily. Water is brackish but during rains it is nearly fresh.

Fresh water swamp forests: These forests grow in low lying areas where rain or swollen river water is collected for some time. Water table is near the surface. Important plants include *Salix tetrasperma, Putranjiva, Holoptelia, Cephalanthus, Barringtonia, Olea, Phoebe, Ficus, Murraya, Adhatoda* and a variety of grasses.

Dry tropical forests

These forests are further classified into three types, such as tropical dry evergreen forests, tropical dry deciduous forests, and tropical thorn forests.

Tropical dry evergreen forests: These forests are found in the areas where rainfall is in plenty but dry season is comparatively longer. The trees are dense, evergreen and short; about 10 to 15 metres high. These forests are found in eastern part of Tamil Nadu, in east and west coasts. The common plant species are much the same as in tropical moist evergreen forests. Bamboos are absent but grasses are common.

Tropical dry deciduous forests: These forests are distributed in the areas where annual rainfall is usually low, ranging between 70 and 100 cm, such as, Punjab, U.P., Bihar, Odisha, M.P. and large part of Indian peninsula. The largest area of the country's forest land is occupied by tropical dry deciduous forests. The dry season is long and most of the trees remain leafless during that season. The forest trees are not dense, 10 to 15 m in height, and undergrowth is abundant. In north, the forests are dominated by *Shorea robusta* and in south by *Tectona grandis*. The other species in south include *Dalbergia, Terminalia, Dillenia, Acacia, Pterospermum, Diospyros, Anogeissus, Boswellia, Bauhinia, Hardwickia, Zizyphus, Moringa, Dendrocalamus*, etc. The other species of northern region are *Terminalia, Semicarpus, Modhuca, Acacia, Sterculia, Launea, Bauhinia, Aegle, Grewia*, etc.

Tropical thorn scrubs: These forests appear in the areas where annual rainfall is between 20–70 cm, dry season is hot and very long. They are found in south Punjab, most part of Rajasthan and part of Gujarat. The vegetation in this region occurs only along the rivers. The land away from the rivers and devoid of irrigation is mostly sandy and devoid of trees. The vegetation is of open type consisting of small trees, 8 to 10 m high and thorny or spiny shrubs of stunted growth. The common species found in these forests are *Acacia, Cassia, Calotropis, Albizzia, Zizyphus, Cordia, Prosopis. Salvadora, Aegle, Atriplex, Grewia, Asparagus, Butea*, etc. The forests remain leafless for most part of the year. There is luxuriant growth of ephemeral herbs and grasses during the rainy season.

Subtropical montane forests

These forests are found in the region of fairly high rainfall but where temperature differences between winter and summer are less marked. Winter generally goes without rains. These are found up to the altitude of about 1500 m in south and up to 1800 m in the north. In composition, subtropical forests are almost intermediate between tropical forests and temperate forests and a sharp demarcation can seldom be made between tropical and subtropical or subtropical and temperate forests. These forests have been grouped into three types such as wet hill broad leaved forests, dry evergreen forests, and pine forests.

Wet hill broad leaved forests: These are found in Karnataka, Assam and M.P. The important species found in the wet hill forests of south are *Eugenia, Terminalia, Murraya, Atylosia, Ficus, Pterocarpus*, etc. while those of the north are *Calamus, Alnus, Quercus, Betula, Cedrella, Garcinia, Populus*, etc.

Dry evergreen forests: These forests appear in the foot-hill areas of Himalayas. The common constituents of vegetation are *Acacia modesta, Olea cuspidata*, etc.

Pine forests: These forests are found mostly in western and central Himalayas and in Assam hills. The forests are dominated by *Pinus khasya, Pinus roxburghii, Quercus, Berberis, Bauhinia,* etc.

Temperate montane forests

These forests occur in the Himalayas at the altitude from 1800–3800 m where humidity and temperature are comparatively low. Montane forests are subdivided into three types on the basis of moisture regime such as montane wet temperate forest, Himalayan moist temperate forest, and Himalayan dry temperate forest

Montane wet temperate forests: These are found in Himalayas extending from Nepal to Assam at the altitude from 1800–3000 m, as well as in the Nilgiris. The forests in south are evergreen. The forests are dense with closed canopy and the trees may be 15–20 m high. Epiphytes are in abundance. Important species in eastern Himalayas include *Hopea, Balanocarpus, Elaeocarpus, Artocarpus, Hardwickia, Salmelia, Dioscoria,* etc. The members of family compositae, rubiaceae, acanthaceae and leguminosae form the undergrowth.

Himalayan moist temperate forests: These forests develop in the areas of lesser rainfall. The trees are high, sometimes up to 45 m tall. The dominant species are oak and conifers. Undergrowth is shrubby and consists of deciduous species of *Barberis, Spiraea, Cotaneaster,* etc.

Himalayan dry temperate forests: These forests dominated by *Rhododendrons,* oaks and conifers from a narrow belt at the altitude from 3000–4000 m in the western Himalayas extending from Uttarakhand through Himachal Pradesh and Punjab to Kashmir. The other commonly found species belong to genera *Daphne, Desmodium, Indigofera,* etc.

Sub-alpine forests

The sub-alpine forests are found throughout Himalayas from Ladakh in the west to Arunachal Pradesh in the east at the altitude from 2800–3800 m. Annual rainfall is less than 65 cm, but snowfall occurs for several weeks in a year. Strong winds and below 0° C temperature prevail for greater part of the year. Trees are like those of temperate zone. Epiphytic mosses and lichens are in abundance.

Alpine forests

Alpine vegetations are classified into three types such as alpine forests, moist alpine scrubs, and dry alpine scrubs.

Alpine forests: Plants growing at the altitude from 2900–6000 m are called alpine plants. In India, alpine flora occurs in Himalayas between 4500 and

6000 m. At lower level, alpine forests consist of dwarf trees with or without conifers and at higher level scrubs and only scattered xerophytic shrubs are left to merge with alpine meadows. The common plants of alpine forests are *Abies, Pinus, Juniperus, Betula, Quercus, Pyrus, Salix*, etc.

Moist alpine scrubs: This type of vegetation is distributed extensively throughout the Himalayas above 3000 m. It is most often dense and composed of evergreen dwarf *Rhododendron* species and other deciduous trees. Mosses and ferns cover the ground with varying amounts of alpine shrubs, flowering herbs and ferns. Alpine pastures include mostly mesophytic herbs with very little grasses.

Dry alpine scrubs: These are open xerophytic formations spread in U.P., Himachal Pradesh, Punjab and Kashmir. Species belonging to *Artemisia, Potentilla, Kochia, Juniperus* predominate in the vegetation which develops generally on lime stone rock.

ROLE OF FORESTS IN FARMING SYSTEMS

Forestry contributes in many ways to sustainable agricultural production and food security. The greatest contribution is through its protective environmental functions such as the maintenance and restoration of soil fertility and soil improvement, erosion control and maintenance of biodiversity. Forestry also contributes in many other ways such as through the direct production of food, provision of rural employment and income.

Maintenance and restoration of soil fertility and soil improvement

Trees improve soils by organic matter maintenance, nitrogen fixation and nutrient recycling. The inclusion of trees in farming systems provide huge quantities of organic matter to the soil through leaf litter, pruning materials and shedding of fine roots. Many tropical tree species fix atmospheric nitrogen to soil. The trees draw nutrients from deeper zone of soil through their extensive root systems. These nutrients come to the upper layer of soil through litter fall. Thus, the nutrients that would otherwise be lost in leaching are recycled. Tree roots reach far down, bringing up water and nutrients from depths that non-woody plants cannot reach. Also, their leaf fall can be used as a natural mulch to increase soil moisture as well as fertility. These beneficial effects of trees have a potential for farming systems. In the absence of fertilizers, soil organic matter supplies a reserve of balanced nutrients, progressively released by mineralization.

Erosion control

Soil erosion is a serious threat to continued agricultural productivity. Erosion whether by wind or water leads to the loss of topsoil where soil nutrients are concentrated thus leading to the disruption of agricultural production and degradation of the soil. Forests and trees reduce wind velocity considerably. Reduction of wind velocity causes considerable reduction in wind erosion. The canopy of trees shelters the ground from the impact of heavy downpours. The leaves drip water on the earth, giving it time to seep underground. Forests increase the infiltration and water-holding capacity of the soil, resulting in much lower surface run-off. This restricts the soil erosion.

Planting trees as windbreaks and shelterbelts reduces the velocity of the wind to a speed that is insufficient to move soil particles. This can also keep seeds and newly germinated seedlings from being blown away or dislodged. The reduction in wind speed leads to lower evaporation from both open water and soil surfaces, making more water available for plant growth.

Maintenance of biodiversity

Maintenance of the biodiversity is an insurance and investment necessary to sustain and improve agriculture. This is because it is the sources of all our food. Moreover, cross breeding of domestic crops with wild varieties can improve yields and produce new strains better adapted to growing conditions or more resistant to diseases and pests. The major store house of this genetic diversity is the forest. The forest systems of the world, particularly tropical forests, house a great number of plant and animal species. Thus, forest ecosystems are very important for both the maintenance and expansion of food production.

Direct production of food

Forests and trees provide food sources in a variety of forms which include edible leaves, fruits, seeds, nuts, roots, tubers, sap, bark, mushroom, honey, snails and insects. Trees are often the only reliable source of food for the family when crops fail. Food from the forests are often used to help meet dietary shortfalls during particular seasons of the year when stored food supplies are dwindling and the next harvest is not yet available. In addition, these products feature prominently during emergency periods such as floods, droughts, famine, wars, economic and social disasters. Besides, a very important source of food is wildlife.

Production of wood and timber

Wood is a major forest produce and is used extensively for various purposes. In India, most of the wood produced is used for construction of houses, bridges, sleepers and wagon, furniture, cabinets, ships, etc. Forests provide raw material to a large number of industries, e.g., paper and pulp, ply board and other boards, saw-mill, furniture, packing cases, matches, toys, etc.

Provision of farm inputs

A variety of farm implements are made from the timbers provided by the forests. A vast variety and amounts of forest and tree products also support the major productive activities of farming including livestock production, fishing and hunting. A shortage of these products constrains the efficiency of crop production. Non-timber forest products (NTFPs) provide materials for supporting crops (e.g., yam and pumpkin stakes), as well as materials for making farm tools. NTFPs also provide materials for making baskets used in carrying and marketing produce, racks for crop drying and storage. Fencing materials are also provided by NTFPs.

Fuelwood supply

Wood is a universal fuel. Forest provides fuelwood needed for the processing of farm produce. For thousands of years, until the advent of coal, oil, gas, electricity, etc., wood constituted man's chief source of fuel. Even today more than half of the total world consumption of wood is for fuel. Wood remains the major source of domestic fuel in India. Supply of firewood from forests releases dung for use, as manure.

As a source of fodder

Many species of trees in the tropics are used for fodder either for browse or stall feeding. Fodder trees contribute in several ways to the overall food security of households. They make a significant contribution to domestic livestock production which in turn influence milk and meat supply. Fodder trees and shrubs have an important advantage over fodder grasses and herbaceous legumes. They can tap deep, underground moisture reserves when the upper soil layers have dried out. Thus, trees can continue to produce fodder when grasses and annual crops have ceased to grow during dry periods.

Protective and ameliorative functions of forest

Forests play a significant role in maintaining the CO_2 balance in the atmosphere. Without sufficient forest cover, all the CO_2 released in the atmosphere will not be utilized, resulting in a higher percent of CO_2 in the atmosphere. This will contribute to global warming. Thus, with more forest cover the effect of global warming can be reduced.

Forests increase local precipitation by about 5–10% due to their orographic and microclimate effect and create conditions favourable for the condensation of clouds. Forests reduce temperature and increase humidity. The effect of forests on temperature is not only limited to forest areas but also it extends far beyond the boundaries of the forests.

Forests check floods. Forests conserve both soil and water. Forests prolong the water cycle from its inception to the final disposal as run-off into streams and ocean. The longer the water retained in the land, the greater is its usefulness in nurturing crops and trees, and in maintaining a regular supply of water in streams throughout the year. Forests increase subsurface run-off which is much slower than surface run-off and the sub-surface run-off does not cause erosion.

Employment and income generation

Forest also contributes indirectly to household food security, through the generation of employment and income from the sale and exchange of gathered and processed forest products. Rural people gather, produce and trade a wide range of forest products in order to derive income. These products include fuelwood, dyes, bamboo, rattan, fibres, fruits, nuts, leaves, mushrooms, medicines, gums, grasses, essential oils, resins, spices, insecticides and lac. These non-wood products are commonly called minor forest products, not because they are of minor significance, but because they are harvested in smaller quantities. In many areas, forestry based activities are a major source of off-farm employment in rural areas.

Chapter 4

Agroforestry System, Subsystem and Practice

The words 'system', 'sub-system' and 'practice' are commonly used in agroforestry literature. In a broad sense, a system is defined as a group of associated elements forming a unified whole and working together in a well defined regular relation for a common goal. An agroforestry system refers to a type of agroforestry land-use that extends over a locality to the extent of forming a land utilization type of the locality. Sub-system and practice are lower-order terms in the hierarchy with lesser magnitudes of role, content and complexity. However, these terms are used loosely, and almost synonymously.

Classification of agroforestry systems is necessary in order to understand and provide a practical framework for evaluating systems and developing action plans for their improvement. Any classification scheme should satisfy the following criteria (Nair, 2008).

1. It should include a logical way of grouping the major factors on which production of the system will depend.
2. It should indicate how the system is managed and find out the possibilities for management interventions to improve the system's efficiency.
3. It should offer flexibility in regrouping the information.
4. It should be easily understood and readily handled.

Different types of agroforestry systems exist in different parts of the world. These systems are highly diverse and complex in character and function. Several criteria can be used to classify and group these agroforestry systems and practices. The most commonly used ones are the system's structure (composition and arrangement of components), its function, its socioeconomic scale and level of management and its ecological spread.

The structure of a system can be defined in terms of its components and their expected roles. This classification considers the composition of the components including spatial admixture of perennial woody component, vertical stratification of the component mix and temporal arrangement of the different components. Based on nature of components, agroforestry systems can be classified into:

- Agrisilvicultural systems
- Silvipastoral systems
- Agrisilvipastoral systems
- Other systems

AGRISILVICULTURAL SYSTEMS

In these systems, annual agricultural crops and perennial woody trees/shrubs are integrated. Based on the nature of the components this system can be grouped into various forms.

i. Shifting cultivation
ii. Improved fallow species in shifting cultivation
iii. *Taungya* system
iv. Multispecies tree gardens
v. Alley cropping
vi. Multipurpose trees and shrubs on farmlands
vii. Crop combinations with plantation crops
viii. Agroforestry fuelwood production
ix. Shade trees for commercial plantation crops
x. Shelterbelts
xi. Windbreaks
xii. Soil conservation hedges
xiii. Rotational woodlots
xiv. Boundary markings

Shifting cultivation

Shifting cultivation is an agricultural land use system in which a piece of forest land is cleared and cultivated until the soil loses fertility. Once the land becomes inadequate for crop production, it is left and allowed to rest longer than the period of cultivation to be reclaimed by natural vegetation. The ecological consequences are often deleterious, but can be partially mitigated if new forests are not invaded. This system is practised extensively in the hill region of north-eastern states and to some extent in hilly areas of Andhra Pradesh, Madhya Pradesh, Bihar, Odisha, Jharkhand, Chhattisgarh and Karnataka. The practice is called '*jhum*' in north-eastern states and '*podu*' in Andhra Pradesh and Odisha.

Demerits of shifting cultivation

- Shifting cultivation is a faulty land use. It causes environmental degradation and ecological imbalance.
- It enhances soil erosion.
- There is soil nutrient loss through run-off, leaching and percolation.
- Soil microbial population is reduced due to cleaning and burning of vegetation.
- The cation exchange capacity of soil is markedly reduced. The carbon:nitrogen ratio is also reduced.
- Shifting cultivation is a primitive method of farming which requires hard labour but gives very less yield.
- Denudation of hill slopes takes place.
- The excessive runoff in the shifting cultivated areas causes floods in the adjoining areas.
- Shifting cultivation disturbs the forest flora and fauna.

Improved fallow species in shifting cultivation

Fallows are croplands left without crops for periods ranging from one season to several years. The main objective of improved fallow species in shifting cultivation is to recover depleted soil nutrients and to check the soil erosion. Once the soil has recovered, crops are reintroduced for one or more seasons. In this system trees and shrubs are not grown with crops on the same plot at the same time. The fallow period varies from region to region, but now it is becoming shorter due to acute land shortage. Mostly the nitrogen fixing tree species are preferred for plantation to improve the soil fertility status. The difference between the improved fallow system and shifting cultivation is that tree species for improved fallows can be selected to fit particular fallow duration in a particular farming system. Although the main functions of the improved fallow species in shifting cultivation are to restore soil fertility and reduce erosion, species choice should not be exclusively confined to 'soil improvers'. Fast growing multipurpose tree species with marketable products should also be selected. Species selected for plantation in the improved fallows should be compatible with the future crops to be raised on the same site, free of any negative physical or chemical effects on the soil.

Taungya system

'*Taungya*' is a Burmese word, consisting '*taung*' means hill and '*ya*' means cultivation *i.e.*, cultivation in the hill. The *taungya* is a modified form of shifting cultivation in which cultivators are required to plant forest tree species and allowed to raise agricultural crops in between rows of tree species until the canopy closes. While shifting cultivation is a sequential system of growing woody species and agricultural crops, *taungya* consists of the simultaneous combination of the two components during the early stages of forest plantation establishment. Although wood production is the ultimate objective in the *taungya* system, the immediate motivation for practising it, as in shifting cultivation, is food production.

Taungya was reported to have started first in Burma (Myanmar) in 1850 by the British for replanting vast areas under teak plantation and in Java in 1856. It was introduced into India by Dietrich Brandis in 1856 and the first *taungya* plantations were raised in 1863 in north Bengal. It is practised in the states of Kerala, West Bengal, Uttar Pradesh, Tamil Nadu, Andhra Pradesh, Odisha, Karnataka and the north-eastern states with an assured annual rainfall of over 1200 mm. A large variety of crops and trees, depending on the soil and climatic conditions are grown in India. The state forest department is the owner of the land and thus the trees. In fact, this system was introduced to raise forest plantation but finally became a recognized agroforestry system. *Taungya* systems are of three types.

Crops grown in taungya system in India

State	Tree crop	Associated agricultural crops
Uttar Pradesh	*Shorea robusta, Tectona grandis, Acacia catechu, Dalbergia sissoo, Eucalyptus* spp., *Populus* spp.	Maize, paddy, sorghum, pigeon pea, soybean, wheat, barley, chick pea, rapeseed
Andhra Pradesh	*Anacardium occidentale, Tectona grandis, Bombax ceiba,* Bamboo, *Eucalyptus* spp.	Hill paddy, groundnut, sweet potato
Kerala	*Tectona grandis, Eucalyptus* spp., *Bombax ceiba*	Paddy, tapioca, ginger, turmeric
Assam	*Shorea robusta*	Paddy
Tamil Nadu	*Tectona grandis,* Bamboo, *Santalum album, Tamarindus indica, Acacia nilotica, Acacia mearnsii, Ceiba pentandra,* Cashew, Rubber	Millet, pulses, groundnut, cotton
West Bengal	*Tectona grandis, Shorea robusta, Schima wallichii, Cryptomeria japonica, Quercus* spp., *Michelia doltsopa*	Paddy, maize, millets, turmeric, ginger, okra, pineapple, sunhemp
Andaman & Nicober Islands	*Pterocarpus dalbergioides*	Sugarcane, maize

Source: Tejwani, 1994

Departmental taungya: In this system, labourers are employed on daily wage basis by the state forest department to grow seasonal agricultural crops along with perennial tree crops. The agricultural crops are raised mainly for suppressing weeds between the tree rows during the early years of plantation.

Leased taungya: The plantation land is leased by the forest department to the interested cultivators for a specified period who raise suitable crops in between the tree rows and take care of tree plantation.

Village taungya: In this system, people are allowed to settle down in a village inside the forest by the forest department and in return, they raise trees and agricultural crops for 3 to 4 years in the land they are allotted. It is the most successful practice of the three *taungya* systems.

Merits of taungya

- Artificial regeneration of forest tree species is cheap.
- Growth of weeds and undesired plants is avoided.
- Employment opportunity is created for unskilled rural people.
- It is a low cost method of establishment of forest plantation.
- It provides food crops from forest land.
- Unutilized forest land is fully utilized by this system.
- It is highly remunerative to the forest department.
- It discourages the practice of shifting cultivation by the tribal.
- This is an important way to integrate rural development programmes.
- It increases the forest wealth of the country.

Demerits of taungya

- The system involves in exploitation of human labour.
- Exposure of land leads to soil erosion and loss of fertility.
- Epidemics are feared due to raising agricultural crops.
- Legal problems may arise due to settlement of villages in forest land.
- Cultivators take interest only on their crops and neglect the trees.
- The villagers may be involved in timber theft and unauthorized hunting.

Multispecies tree gardens

This system consists of a mixture of tree plantations of conventional forest species and other commercial perennial tree crops that gives a managed mixed forest appearance. As opposed to homegardens which surround individual houses, these tree gardens are usually away from houses, and are typically found on community owned lands surrounding villages with dense clusters of houses. The multispecies, multilayer dense plant associations are with no organized planting arrangements. The major groups include different woody components of varying forms and growth habits. Perennial woody fruit trees are also included in the system. Herbaceous plants are usually absent, but the shade tolerant ones may sometimes be present. The major function of this system is production of food, fodder and wood products. This agroforestry practice is adaptable to areas with fertile soils, with good availability of labour and high human population pressure. Important woody species planted in this system are *Acacia catechu, Phoenix dactifera, Artocarpus* spp., *Acacia mangium, Acacia auriculiformis, Gmelina arborea, Mangifera indica, Syzygium aromaticum*, etc.

Alley cropping

Alley cropping or hedgerow intercropping is a management-intensive agroforestry practice in which perennial, preferably leguminous trees or shrubs are grown simultaneously with an arable crop. The trees, managed as hedgerows, are grown in wide rows and the crop is planted in the interspaces or 'alley' between the tree rows. The trees are pruned regularly during the cropping phase and allowed to grow freely to shade the inter-rows when there are no crops. Alley cropping retains the basic restorative attributes of the bush fallow through nutrient recycling, fertility regeneration and weeds suppression and combines these with arable cropping so that all processes occur concurrently on the same land, allowing the farmer to crop the land for an extended period.

Benefits from alley cropping

- An important benefit of alley cropping is the addition of large amounts of organic materials from the prunings as mulch or green manure which can have favourable effects on soil physical and chemical properties, on microbiological activity and hence on soil productivity. This ultimately improves crop performance in alleys.
- There is a reduction in the use of chemical fertilizers which decreases environmental pollution and maintains soil health.
- There is an overall improvement in the physical nature of the soil environment. The addition of mulch moderates soil temperature, reduces evaporation and improves activity of soil fauna and soil structure resulting in better infiltration, reduced runoff and improved water use efficiency.
- The tree rows on sloping land act as a physical barrier to soil and water movement, resulting in significant reductions in erosion losses. The presence of prunings applied as mulch in the alleys also controls soil erosion.

- Alley cropping provides additional products such as forage, firewood or stakes when a multipurpose tree is used as the hedgerow.
- During the fallow period shading of the interspaces reduces weed growth, while in the cropping phase, the mulch inhibits germination and growth of weeds.

Management of alley cropping systems

The effectiveness of an alley cropping system depends to some extent on the soil type and agroecological zone in which the system is practised. But the success of the system is also very dependent on management strategies adopted. Factors such as choice of tree species, orientation, layout and manipulation of the hedgerows and crop husbandry practices are all important in determining the outcome of the alley cropping system. Trees are planted as hedgerows in farm fields to maximize the positive and minimize the negative effects of trees on crops. Trees compete with agricultural crops for nutrients, soil moisture and solar energy. However, right kind of tree species planted at right spacing, with proper management practices can reduce competition and actually produce a net increase in yields per unit area.

The position and spacing of hedgerow and crop plants in an alley cropping system depend on plant species, climate, slope, soil condition and the space required for the movement of people and tillage equipments. Ideally, hedgerows should be positioned in an east-west direction so that plants on both sides receive full sunlight. The spacing adopted is usually 2 m to 8 m between rows and 25 cm to 2 m between the trees within rows. The closer spacing is generally used in humid areas and the wider spacing in subhumid and semiarid regions. On sloping land, hedgerows are densely planted along the contours to form a barrier against soil erosion. Grass strips planted alongside hedgerows will create a more effective barrier. If the desirable east-west orientation of hedgerows is not maintained on the slopes then regular pruning of tree branches is required to prevent excessive shading to the agricultural crops in alleys.

The choice of tree species for alley cropping is extremely important and to a large extent determines the success or failure of the system. The following attributes are to be considered when selecting a tree species for an ideal alley cropping.

- rapid growth rate
- ability to withstand frequent cutting
- good coppicing ability (regrowth after cutting)
- easy to establish from seeds or cuttings
- nitrogen fixing capacity

- deep-rooted with a different root distribution to the crop
- multiple uses such as forage and firewood
- ability to withstand environmental stresses such as drought, waterlogging and extremes of pH
- high leaf to stem ratio
- small leaves or leaflets
- dry season leaf retention
- freedom from pests and diseases

The suitable species for hedgerow planting are *Cassia siamea, Leucaena leucocephala, Glyricidia sepium, Calliandra calothyrsus, Sesbania sesban, Acacia mangium, Gmelina arborea, Albizia* spp., *Cajanus* spp., *Dalbergia sissoo, Tectona grandis, Chamaecytisus* spp., *Desmodium* spp., *Erythrina* spp., *Flemingia* spp., *Inga* spp. and *Tephrosia* spp.

Multipurpose trees and shrubs on farmlands

In this system, trees are scattered haphazardly or according to some systematic patterns on bunds, terraces or plot/field boundaries, crop lands, pastures and rangelands. Some of the trees grow naturally from seed dispersed by birds and other wildlife. Such trees are retained by farmers during land preparation for agriculture and are often randomly dispersed on the land. The most important and dominant tree species in this system are *Acacia albida, Casuarina equsetifolia, Leucaena leucocephale, Vitellaria paradoxa, Parkia biglobosa, Adansonia digitata, Tamarindus indica, Acacia tortilis, Azadirachta indica, Acacia senegal, Cocos nucifera, Pongamia* spp. and *Prosopis* spp.

Benefits

- Farmers consider such trees not to be competitive against food crops.

  Trees provide shade for livestock during intense heat of the long dry season.
- The trees diversify farmers' products and increase crop production and the duration of cropping season without fertilizer use.
- The system reduces soil erosion due to high wind velocity or rain water runoff.
- Sale of non-timber products, such as charcoal, firewood, tree borne oil and fruits significantly increases income.

Limitations

- Scattered trees are often not at the optimum density that would provide maximum benefits to the environment and crop production.
- Reliance on naturally regenerating trees makes it difficult to improve the system through use of better germplasm.

- Some self regenerated tree species are slow growing and so, benefits take long to accrue.
- The trees are often browsed by livestock that are allowed to graze crop residues during the dry seasons.

Crop combinations with plantation crops

On large estates plantation crops such as tea, coffee, cashew, oil palm, rubber, cocoa and coconuts are usually grown in monoculture. But intercropping in these perennial tree crops on small holdings is possible or even desirable due to several reasons.

1. There are large spaces between tree crops during the early stages of growth.
2. The intercrops reduce soil erosion between widely spaced tree crops, especially during the early tree growth stages.
3. Roots of some tree intercrop combinations complement each other.
4. Shade tolerance of understorey crops such as arrowroot, pineapple, turmeric, ginger, *Aloe vera*, etc. favours growth under some plantation crops.
5. Complementary use of light *e.g.*, up to 8 years and after about 25 years, coconut palms allow a considerable amount of light to reach the intercrops. On the other hand, while oil palm has a high light requirement that is available in the upperstorey, cocoa in the understorey requires considerable shade, except at flowering stage.
6. Intercropping during early growth period of tree crops is economically viable while waiting for harvest of long maturity tree crops.
7. Some intercrops are compatible because of differing peak labour requirement time with that of plantation crops.
8. Some tree crops require an intercrop to serve as a nurse crop during the early growth period. For instance, newly transplanted coffee seedlings benefit from the shade provided by taller intercrops, such as maize.

Agroforestry for fuelwood production

In this system, various multipurpose fuelwood species are interplanted in or around agricultural lands. The primary productive role of this system is to produce firewood. Tree species commonly used as fuelwood are *Acacia nilotica, Acacia auriculiformis, Acacia mangium, Gmelina arborea, Dalbergia sissoo, Albizia lebbeck, Casuarina equisetifolia, Prosopis juliflora, Cassia siamea, Eucalyptus tereticornis*, etc.

Shade trees for commercial plantation crops

Trees are often planted to provide shade for plantation crops such as coffee and tea. Young cocoa plants also need shade in the nursery and for the first 2-3 years in the field. The shade moderates the microenvironment so that excessive moisture stress is avoided during dry season. The use of shade trees in coffee plantations is common in the tropics and legume trees are mostly used for this purpose. In addition to providing shade, these trees contribute nutrients to low fertility soils where inorganic fertilizers are not used. The important shade providing tree species are *Erythrina poeppigiana, Cordia alliodora, Gliricidia sepium, Leucaena leucocephala, Albizia chinensis, Grevillea robusta*, etc.

Shelterbelts

A shelterbelt is a wide strip of vegetation consisting of several rows of trees that slows wind speed, thereby reducing wind erosion, evaporation and damage of field crops by the wind. It is sometimes referred to as windbreak, although the latter often implies a single strip of trees and other vegetation. A shelterbelt presents a mechanical barrier to the impact of the wind and separates two zones, the windward and the leeward zones. The windward zone refers to the side from which the wind blows, while the leeward zone relates to the side where the wind passes. The main characteristics of shelterbelts are

1. Shelterbelts have a typical triangular shape acquired by raising tall trees in the centre.
2. A certain degree of penetration by wind is planned because by raising a solid wall of trees, the protection decreases very fast on the leeward side.
3. The ratio of height and width of a shelterbelt should be roughly 1:10. Shelterbelts up to 50 m width are considered ideal under Indian conditions.
4. Orientation of shelterbelts depends upon the direction and velocity of the prevailing winds. Rows of trees are established at right angles to the prevailing wind.
5. The minimum length of a shelterbelt should be about 25 m.
6. Tree species selected for shelterbelt establishment should have ability for wider adaptability to climatic conditions (drought, frost, extreme temperatures, etc.), high growth rate, quick crown formation, developing a strong and deep root system (so that they do not compete for moisture and nutrients with agricultural crops), evergreen foliage, increased land productivity.

Benefits from shelterbelts

- **Reduces soil erosion by wind:** A shelterbelt modifies the microclimate, mostly in its leeward side. This modified microclimate includes reduced wind speed and, therefore, reduced soil erosion. A significant reduction of wind speed occurs on leeward side for a distance extending to approximately 20 times the height of the shelterbelt and also 3 to 5 times its height on the windward side. Therefore, a shelterbelt 5 m in height will provide a degree of protection for soils and crops for a total distance of up to 25 times its height *i.e.*, 125 m. Since the zone of protection provided by a single shelterbelt is limited, a series of shelterbelts is required to protect the whole field. The shelterbelts must be planted perpendicular to the direction of the prevailing wind to provide complete protection.

- **Increases moisture for crop growth:** Shelterbelts reduces evaporation, thus providing more moisture for crop growth. It also protects the leeward areas from the desiccating effects of hot wind. Shelterbelts use moisture and nutrients from a greater depth than most annual crops.

- **Reduced wind damage to crops:** Crops benefit from the reduced wind speed in the protected zone. The crop plants are less likely to be twisted by the wind.

- **Potential for increased farm productions:** Most of the research conducted around the world reports crop yield increases due to shelterbelts. However, crops vary in their yield response to shelterbelt protection. Drought tolerant crops show the lowest response, forage crops are moderately responsive and weather-sensitive crops such as vegetables show the highest response. Besides, shelterbelts provide potential source of income for farmers (*e.g.*, biomass, timber and non-timber products).

Windbreaks

Windbreaks are narrow strips of trees, shrubs and/or grasses planted to protect crop fields, homes, canals and areas from wind and blowing sand. Where wind is a major cause of soil erosion and moisture loss, windbreaks can make a significant contribution to sustainable production. From the farmers' point of view, gentle winds are advantageous as they can contribute to pollination of crops and seed dispersal. Strong winds, on the other hand, are damaging and could be detrimental to agricultural crops, human life and properties. Below are some of the effects of strong winds on crops.

- Wind may increase transpiration rate, and this may lead to soil moisture deficits.

- It can spread pests and diseases because disease causing spores and insects are dropped whenever wind speed is reduced.
- It can cause crop lodging and deform plants.
- It can increase loss of topsoil through wind erosion especially in semiarid areas.
- It can cause drifts during herbicide and insecticide spray, hence, leading to wasteful application and ecotoxicity.

Benefits from windbreaks

- It protects crops and pastures from hot and dry winds, cold winds, and frost.
- It reduces/prevents soil erosion.
- It improves the microclimate by moderating temperature.
- It reduces evaporation from farm lands, also checks transpiration.
- It prevents grass fire.
- It provides fencing and boundary demarcation.
- It increases overall farm productivity.

Design and management

The basic design of a windbreak should include permeability, the appropriate orientation, placement, length and height, number of rows, spacing, density and continuity to provide effective protection.

Permeability: The effectiveness of the windbreak is influenced by its permeability. If it is dense, like a solid wall, the airflow will pass over the top of it and cause turbulence on the leeward side due to the lower pressure on that side. This gives a comparatively limited zone of effective protection on the leeward side compared to the zone that a moderately permeable shelter creates. Windbreaks must be semi-permeable, ideally filtering 50–60 percent of the wind to reduce its strength. The desired permeability can be obtained by carefully selecting tree species. Species such as *Eucalyptus* and *Casuarina* will form effective windbreaks. Permeability of dense windbreaks can be improved by pruning lower branches at 0.5-0.8 m from the soil level.

Orientation: When designing a windbreak, the direction of the wind must be considered. A barrier should be established perpendicular to the direction of the prevailing wind for maximum effect. To protect large areas, a number of separate barriers can be created as parts of an overall system. When the prevailing winds are mainly in one direction, a series of parallel shelterbelts perpendicular to that direction should be established. Before establishing windbreaks or shelterbelts, it is important to make a thorough study of the local winds and to plot on a map the direction and strength of the winds.

Height and length: While selecting species for windbreak more importance is given to the height than the thickness. It is generally accepted that a windbreak protects an area over a distance up to 3-5 times its own height on the windward side and up to 20 times its height on the leeward side. Thus, the species selected should be erect and tall growing, hardy and drought resistant, and should occupy less space as far as possible. Windbreaks are more effective when they stretch without major gaps for distances exceeding 12 times the mature height of the trees.

Tree rows and spacing: In reducing wind speeds, narrow barriers can be as effective as wide ones. Although in theory, one-row barrier should suffice, in practice three to five rows are more effective. If a single row windbreak is to be planted, tree species that retain their foliage to the ground and give a fairly dense growth should be selected. *Eucalypltus* tree are generally unsuitable as single row windbreaks because of their habit of losing the lower branches. The main disadvantage of a single row is that if one tree is lost, gap is created which reduces the efficiency of the entire windbreak. Distance between trees varies with the relative importance of the protective versus productive purposes of the windbreak. Initial spacing of 3 meters between the rows, with trees 1–2 meters apart in the rows is desirable.

Species: In general, tree or shrub species selected for windbreaks should have the characteristics of rapid growth, straight stems, wind firmness, good crown formation, deep taproot system and resistance to drought. Some of the tree genera used for windbreaks are *Cassia, Acacia, Casuarina, Leucaena, Grevillea, Cupressus, Pinus, Dalbergia, Syzygium, Erythrina, Mangifera, Bambusa,* etc.

Windbreak for fruit orchards
Fruit orchards when exposed to strong wind usually incur heavy losses. Heavy wind increases the losses of moisture both by increasing transpiration and surface evaporation. The high winds also cause damage to fruit trees by breaking of branches, destruction of blooms and dropping of immature fruits. The growth and yield in protected orchard is definitely better than the exposed orchard. Hence, establishment of a tall growing windbreak is necessary to protect an orchard. Planting of windbreaks should be done at least two years after planting of fruit trees. The first row of it should be planted 10 m away from fruit plants at a close spacing to form a thick screen. The windbreak trees sometimes may compete with the fruit trees for water and nutrient. To prevent this competition, a trench may be dug about 1 m deep and 5 m away from the row of windbreak trees and all the roots of windbreak trees exposed in the trench are cut off periodically.

Multipurpose windbreaks

Although planting windbreaks is an investment that takes some land out of production, well designed windbreaks protect the health and productivity of crops enough to make the overall return positive. A multipurpose windbreak is designed to provide multiple functions and/or products, in addition to wind protection. Multiple products from a windbreak can include yields such as fruit, timber, animal fodder, mulch, wildlife habitat and other economic farm products. Adding multiple functions or products to a windbreak plan can make the installation and management more satisfying and economically viable for the farmer. Multipurpose windbreaks require special care in planning and management to maintain the primary function of wind protection while maximizing secondary yields. Once the form and position are carefully determined, then multiple functions or products can be added.

Since planting trees for a windbreak involves a long term investment, trees having timber value may also be included. The main drawback of having timber as a secondary yield from a windbreak is that wind damage may produce timber of poor quality. Also, since windbreak trees require no or very little pruning, the lack of pruning may reduce timber quality and quantity in certain species that require a lot of pruning for optimal timber production. Planning for timber harvest requires the most careful effort as the entire trees are removed. The planting, harvesting and replanting must be coordinated to avoid the creation of gaps in the tree rows. Integrating timber trees with permanent rows of non-timber windbreak trees will help maintain the effectiveness of the windbreak. *Grevillea robusta, Pterocarpus indicus* and *Azadirachta indica* are some of the best suitable species for this purpose.

Nitrogen fixing trees (NFTs) can also be integrated in a multi-row windbreak and pruned regularly to provide a nutrient-rich mulch for crops, or a nutritious fodder to supplement the diet of farm animals. Although pruning should be avoided for most windbreak trees, the practice of cutting back NFTs and allowing them to resprout can be integrated with windbreak management. Pruned NFTs are much more susceptible to wind damage if they are allowed to regrow to a large size, but if they are cut regularly and the regrowth kept small they will be effective as a short row. To maintain the windbreak's primary function with this practice, it is essential to prune the NFTs regularly. These species are planted on the most sheltered side of the windbreak. The important species for this purpose are *Leucaena leucocephala, Sesbania sesban, Calliandra calothyrsus,* and *Gliricidia sepium.*

Guidelines for multipurpose windbreak design

1. The species used should be selected first for their wind tolerance and appropriateness for the site (climate, soils, etc). The products should be a secondary consideration in selecting species.
2. Windbreaks designed for multiple products should comprise of multiple rows. It enables more flexibility in management and harvest of products without compromising wind protection by creating gaps.
3. Trees yielding products such as fruit, food, fodder or mulch should ideally be located in the interior or wind-sheltered rows of the windbreak for maximum protection.
4. A diversity of species should be used to allow for greater flexibility in management and for better resistance to damage to trees caused by insects or diseases.

Soil conservation hedges

Trees and shrubs are planted on physical soil conservation works such as grass strips, bunds and terraces wherein they play two roles *viz.* (i) to stabilize the soil conservation structure through the root system and (ii) to make productive use of the land they occupy. The risers or terraces in steeply sloping landscapes are densely planted with multipurpose and/or fruit trees. The primary role of multipurpose/fruit trees and agricultural species is soil conservation and provision of various tree products. The important tree species used for soil conservation include *Grevillea robusta, Acacia catechu, Pinus roxburghii, Acacia modesta, Prosopis juliflora, Alnus nepalensis, Leucaena leucocephala*, etc.

Rotational woodlots

In rotational woodlots, multipurpose trees and shrubs are intercropped with food crops for about 2–3 years, when the trees are still too small to out-compete the food crops. This phase resembles the *taungya* system. In the subsequent phase, the trees are allowed to grow to maturity in pure stand. Therefore, the second phase resembles a fallow period. Soil fertility may be improved if the trees fix nitrogen and litter fall add large amounts of organic matter to the soil. When the trees are harvested, crop production resumes and the crops benefit from the residual soil fertility built-up during the 'fallow' period. The main function of rotational woodlots is to produce both food crops and tree products and services such as fuelwood, poles, fodder and soil fertility improvement. Because a woodlot keeps the soil surface covered, it is also a soil and water conservation measure. *Acacia auriculiformis, Azadirachta indica, Dalbergia sissoo, Albizia lebbeck, Gmelina arborea, Delonix regia,*

Anogeissus leiocarpus, *Cassia fistula*, etc. are all excellent tree species for a woodlot.

Boundary markings

Certain multipurpose trees are used to mark farm boundaries. Unlike live fences, the trees on boundaries need not be closely spaced except when soil erosion control is also desired. Tree growing on farm boundaries requires agreement between neighbouring farm owners to avoid conflicts. Sometimes two rows of trees are planted, one on each side of the boundary, and then each farmer grows and manages his own trees. A disadvantage of this is that it occupies more land than a single row. Initially trees can be established at a close spacing (0.75–1.00 m) and then later thinned for poles or firewood to a final spacing of 1.5–3.0 m. With double rows the spacing between the rows should not be less than 2 m.

SILVIPASTORAL SYSTEMS

Silvipasture as an agroforestry practice is specifically designed and managed for the production of trees, tree products, forage and livestock. Silvipasture results when forage crops are deliberately introduced or enhanced in a timber production system, or timber crops are deliberately introduced or enhanced in a forage production system. Integrating livestock production in the farming system promotes farm product diversification, improves food security and offers a diversified marketing opportunity that can stimulate rural economic development. The interactions among timber, forage and livestock are managed intensively to simultaneously produce timber commodities, a high quality forage resource and efficient livestock production. Overall silvipastoral systems can provide economic returns while creating a sustainable system with many environmental benefits.

Before new silvipastoral systems are established, implications of merging forestry and agricultural systems should be explored thoroughly for economic and environmental considerations. Factors like availability of potential markets, soil type, climatic conditions and species compatibility should be considered while choosing tree and forage crop species. The timber component should be marketable, of high quality, fast growing, deep rooted, drought tolerant, and capable of providing the desired products and environmental services. The forage component should be a perennial crop that is suitable for livestock grazing, compatible with the site (soil, temperature, precipitation), productive under partial shade and moisture stress, responsive to intensive management and tolerant of heavy utilization or grazing. A successful silvipastoral system requires understanding forage growth characteristics and managing the timing and duration of grazing to avoid browsing of young tree seedlings. Improper management of silvipastures can reduce desirable woody and herbaceous plants by overgrazing and soil compaction.

Benefits from silvipastoral systems

1. Economic risk is reduced because the system produces multiple products, most of which have an established market. Production costs are reduced and comprehensive land utilization in silvipastoral systems provides a relatively constant income from livestock sale and selective sale of trees and timber products. Well managed forage production provides improved nutrition for livestock growth and production.
2. Grazing can enhance tree growth by controlling competition of grass for moisture, nutrients and sunlight. Well managed grazing provides economical control of weeds. Fertilizer applied for forage is also used by trees. In addition, livestock manure recycles nutrients to trees and forage.
3. Some forage species tend to be lower in fibre and more digestible when grown in a tree-protected environment. Trees that provide shade or wind protection can have a climate-stabilizing effect to reduce heat stress on livestock. Protection by trees can cut the direct cold effect by 50 percent or more and reduce wind velocity by as much as 70 percent.
4. The forage protects the soil from water and wind erosion, while adding organic matter to improve soil properties. Some forage species fix nitrogen.
5. Silvipastures provide an attractive landscape with an aesthetic value. It can also increase wildlife diversity.

Within the broad category of silvipastoral system there are several types of practices which can be identified depending on the role of the tree/shrub (sometimes collectively called 'trub') component. These practices include:
  i. Protein banks (fodder tree banks)
  ii. Trees and shrubs on rangeland or pastures
  iii. Live fences of fodder trees and shrubs
  iv. Plantation crops with pastures and animals

Protein bank

Protein banks are blocks of forage plants deliberately planted to alleviate fodder shortages in arid, semiarid and mountainous regions, especially during the dry seasons. The forage plants may be leguminous trees and shrubs or herbaceous legumes, and they may be grown in combination with suitable grasses. The fodder trees are pruned regularly to feed livestock. When based on legumes, the fodder banks become important sources of protein and are referred to as protein banks. The important tree and shrub species for this system are *Acacia nilotica, Albizia lebbeck, Leucaena leucocephala, Gliricidia sepium, Sesbania grandiflora, Artocarpus* spp., *Bombax malabaricum, Cordia dichtotoma*, etc.

Trees and shrubs on rangelands or pastures

The primary use of trees in range and pasture land is to provide shade. Besides this vital function, trees on rangeland also provide other benefits like fodder and wood. Normally such trees are scattered at random but sometimes they may be arranged according to a systematic pattern. The number of species with potential for this practice is great. They include *Acacia nilotica, Acacia mangium, Acacia auriculiformis, Gliricidia sepium, Bauhinia* spp., *Leucaena leucocephala, Derris indica, Azadirachta indica, Emblica officinalis, Psidium guajava, Prosopis* spp. and *Tamarindus indica*.

Live fences of fodder trees and shrubs

Live fences are lines of trees or shrubs planted on farm boundaries or on the borders of farmyards, pasture plots or animal enclosures. Sometimes they are also used around agricultural fields. They serve mainly as field boundaries to protect the property from stray animals. They can be made of densely planted single or multiple rows. The shrubs or trees are regularly pollarded and trimmed. The live fences provide shade, protection and privacy for the livestock integrated in this silvipastoral system. The trees can also serve as windbreaks that produce wood and foliage products. The foliage can be used as fodder. Legumes are especially valued as they have high protein content. In arid and semiarid zones, live fences are often made of thorny species of the *Acacia* or *Prosopis* genera. In the humid and subhumid tropics, leguminous species such as *Gliricidia sepium, Sesbania grandiflora* and *Erythrina berteroana* are used.

Plantation crops with pastures and animals

Plantation trees such as coconut, cashew, mango, etc. can be planted in scattered form on pasture land. Cattle can graze in open and lightly shaded pastures. This practice is beneficial to the animals in many ways.

- Pastures may grow more as trees bring up nutrients from below pasture rooting depth.
- Trees may improve quantity and quality (*e.g.*, protein, minerals, energy) of forage available.
- Light shade may help pasture growth and quality in dry areas by improving soil surface microclimate.

- Trees may supply fodder when pasture species are infested by insects or diseases.
- Fodder tree leaves are less trampled than those of creeping or erect pasture grasses and legumes.
- By having shade, animals for meat grow better and dairy cows produce more milk in the humid tropics.
- Wind and cyclone protection is provided by the trees.
- Animals can eat tree fruits and pods which are rich in minerals and vitamins.

AGRISILVIPASTORAL SYSTEM

Agrisilvipastoral system is the combination of agricultural crops, woody perennials and livestock. Agroforestry practices under this system include
  i. Agrisilvicultural system converted to silvipastoral systems
  ii. Multistorey system with free grazing
  iii. Homegardens (tree-livestock-crop mix around homesteads)
  iv. Woody hedgerows for browse, green manure, soil conservation, etc.
  v. Alley cropping system using pasture grasses/fodder crops and agricultural crops

Agrisilvicultural system converted to silvipastoral systems

In this system, the initial cropping combinations include tree seedlings and annual agricultural crops. As the trees grow and close their canopies, the yield recovery of annual agricultural crops gradually decreases and it will no longer be possible to grow these crops. Instead, shade tolerant grasses and vines will be more compatible to land use system where animals are allowed to graze freely, thus converting agrisilvicultural system to silvipastoral system.

Multistorey system with free grazing

This is similar to the multistorey under agrisilvicultural system, except that in this case, grazing animals are an added component.

Alley cropping with pasture grasses and agricultural crops

This is similar to hedgerow cropping with seasonal agricultural crops. However, instead of all alleys planted with agricultural crops, some alleys in between the hedgerows are grown with improved pasture grasses and/or fodder trees or shrubs which are regularly cut and fed to livestock.

Woody hedgerows

In this system, various woody hedges especially fast growing and coppicing fodder shrubs and trees are planted for the purpose of browse, mulch, green manure, soil conservation, etc. The species for this system are *Erythrina* spp., *Leucaena luecocephala*, *Sesbania grandiflora*, etc.

Homegardens

This is one of the oldest agroforestry practices found extensively in the high rainfall areas in tropical south and south-east Asia. The word 'homegarden' has been used to describe diverse practices from growing vegetables behind houses to complex multistoreyed systems. It is the intimate association of multipurpose trees and shrubs with seasonal and permanent arable crops and livestock within the compounds of individual houses. Mostly the whole system *i.e.*, crop-tree-animal unit is managed by family labour. These systems are common in all ecological regions in the tropics and subtropics, especially in humid lowlands with high population density. The average size of a homegarden is usually much less than one hectare. In India, every homestead has around 0.2 to 0.5 ha land for personal production. In many parts of the world the fruit, vegetable and other foodstuff produced from homegardens provide a substantial part of the household food requirement. Food production is the primary function and role of most of the homegardens. The combination of crops with different production cycles results in a relatively uninterrupted supply of food products throughout the year. In most cases, animals and birds are kept in the homegardens. Fodder and legumes are widely grown to meet the daily fodder requirement of the animals. The waste materials from crops and homes are also used as fodder for animals and feed for birds. Cattle are mostly kept for dairy products and land cultivation. Goats, chickens and fish are kept for household consumption. Products from animals or the animals themselves can also be sold in the market.

Structure of homegardens

The layered configurations and combination of compatible species are the most important characteristics of all homegardens. The gardens are carefully structured systems where each component has a specific place and function. Every homegarden usually consists of an herbaceous layer near the ground, a tree layer at the upper level, and intermediate layers in between. The lower layer is usually divided into two, with the lowermost (less than 1 m height) dominated by different vegetable and medicinal plants, and the second layer (1-3 m height) being composed of food plants such as banana, papaya, yam and so on. The upper tree layer is also partitioned into two, consisting of emergent, fully grown timber and fruit trees occupying the uppermost layer of over 25 m height, and medium-sized trees of 10-20 m height occupying the next lower layer. The intermediate layer of 3-10 m height is dominated by various fruit trees.

Benefits of homegardens

- Production is very diverse and continuous with less risk and there is a regular flow of products such as food, fuel, fodder, fruit, spices and timber.
- The protective function of homegardens is very high with soil conservation, water retention and favourable microclimate for humans and animals.
- The use of labour for the farm is very efficient due to proximity of the garden to the house, within the walking distances.
- Valuable crops or animals in need of protection can be given extra care by the farm family because of the short distance from the house.
- Homegardens ensure a pleasant living environment, shade and provision of privacy.
- They can also provide aesthetic values and open patches in the homegarden often serve as family or village gathering places.

Constraints of homegardens

- The high diversity of ecological niches in a homegarden may provide a habitat for species that can become pests like snakes, insects and fungi.
- During tree harvesting other plants are often damaged.
- Proximity to homesteads makes houses prone to damage from falling branches and extensive lateral roots from trees.

OTHER AGROFORESTRY SYSTEMS

Other minor agroforestry technologies include aquaforestry, agrisilviaquaculture, apisilviculture, sericulture, etc.

Aquaforestry

The aquaforestry system comprises of composite fish culture in farm ponds, and various trees and shrubs (*Leucaena leucocephala, Morus alba, Gliricidia sepium, Moringa olifera*, etc.), leaves of which are preferred by fish are planted on the boundary and around fish ponds. Leaves of these trees are used as feed for fish. Inland fish such as *catla, rohu, mrigal*, common carp, silver carp and grass carp can be grown in the ponds. In the coastal regions farmers are cultivating fish and prawn in saline water and growing coconut and other trees on bunds of ponds. Now fish culture in the mangroves is also advocated, which form a rich source of nutrition to the aquatic life and breeding ground for fish and prawn. Other terms that have been used for this practice are silvipisciculture, agrisilviaquaculture and aquasilviculture. Some other modified aquaforestry systems are

1. A well-balanced system of animal husbandry including goatery, poultry, duck farming, turtles and fishes in the small ponds in homegardens make a balanced system of high moisture, energy and nutrient use efficiency per unit area.
2. In paddy field, fish can easily be reared by planting trees on field bunds or boundary to provide leaves used as fish feed. This system can be practised in high rainfall areas.
3. Coconut plants can also be successfully planted on raised paddy field bunds with an alley space of 5 m width depression which is utilized for pisiculture.

Apisilviculture

In this system, various honey or nectar producing trees frequently visited by honeybees are planted on the boundary of the agricultural field. Bees benefit the trees and trees in turn provide series of benefits to bees. The primary purpose of this system is to produce honey. Api-silviculture with *Eucalyptus, Gliricidia, Grevillea, Gmelina, Leuceana* and *Albizia* species were more remunerative and a good source of generating additional farm income in rural areas. Fruit trees can also be integrated with apisilviculture for more profits.

Sericulture

It is the culture of silkworms. Mostly, these silkworms are fed with leaves of mulberry tree (*Morus alba*). Silkworms produce silk and several by-products, which are used for many purposes like textile, fibre, soap, vitamins, medicine, etc. Silkworms depend upon mulberry tree leaves for their normal growth and development. The mulberry tree is a small deciduous tree which tolerates drought and heat but prefers moist climate. The mulberry tree also gives a fine wood. Its branches can be used to make farm tools and its bark to make high quality paper and artificial fibre. The fruit is edible and can be used to make wine. The tree can produce fuelwood and fodder. It also provides shade and can be used as ornamental plant. It is useful for windbreaks, soil conservation, soil stabilization, hedgerow planting, bee forage and live fence.

Mushroom in mixed tree species

Cultivation of paddy straw mushroom under the shade of high density plantation of mixed multipurpose tree species is also remunerative. The tree species suitable for the system are *Acacia auriculiformis, Acacia mangium, Casuarina equisetifolia, Gmelina arborea, Dalbergia sissoo*, etc.

Multipurpose woodlots

In this system, location specific multipurpose tree species are grown mixed or separately for various purposes such as wood, fodder, soil protection, soil reclamation, etc. Although not considered an agroforestry practice *per se*, short-rotation intensive silviculture offers definite potential for the reclamation of abandoned land and is therefore one of the technological options that should be considered on agricultural land in harmony with other types of agroforestry production systems.

Chapter 5

Tree Crop Interaction in Agroforestry

Agroforestry systems are not simply systems where trees and crops or animals give useful products to the farmers, rather systems where trees and crops and/or animals interact. Interaction literally means influence or mutual or reciprocal action. So component interaction refers to the influence of one component of a system on the performance of other component as well as the system as a whole.

In agroforestry systems, trees are grown in close proximity to crops and pasture. Their performance would largely depend on their ability to share various growth resources in a given environmental situation. Various interactions take place between the woody trees and herbaceous plants (crops or pastures) which is referred to as tree-crop interface. These interactions take place through the media of soil and microclimate and may exert favourable or adverse effects on the crop. Study of interaction helps to know how the components of agroforestry utilize and share the resources of the environment, and how the growth and development of any of the components will influence the others. Interaction occurs both above and below the ground and includes a complex set of interactions relating to radiation exchange, the water balance, nutrient budget and cycling, shelter and other microclimatic modifications.

The success of an agroforestry system relies heavily on exploitation of the component interactions. In an ideal relationship, production of trees as well as crops or grasses in combination could be comparable to their sole performance. Agroforestry could be even more advantageous if the production of associated components is increased due to influence of trees. This is possible because trees are capable of improving productivity of soil in many ways. A large number of trees are known to fix nitrogen symbiotically. Nevertheless, instances of crop inhibition in association of trees are not uncommon. Such inhibitions are primarily caused by shade effect as well as competition for below-ground resources such as nutrients and water. In some cases inhibitory effect may also result from allelochemicals secreted by some of the tree species.

NATURE OF INTERACTIONS

The nature of interactions between two components can be described on the basis of observable net effect of one component on another in a system. The basic principle deciding the nature of interactions depends upon the ability of different components to capture and use the most limiting essential growth resources effectively. The capture of the limiting resource (*e.g.*, light, water or nutrients) depends on the number, surface area, distribution and effectiveness of the individual elements within the canopy or root system of the species or mixture of species involved.

Complementary interaction

In a system if the tree and crop components help each other, by creating favourable conditions for their growth in such a way that the agroforestry system provides a greater yield than the yield of their corresponding sole crops, then the interaction between the components is said to be complementary in nature. The complementary interaction is of two types, *viz.*, spatial and temporal. When complementarity is achieved simply by varying the proportion of the desired species, it is known as spatial complementary interaction. In this, the geometry of the system is altered by changing the spacing and density of the components. Therefore, under complementary association the proportion of each species can be modified to suit one's requirements without any sacrifice. When two species complement each other, the capture of the major limiting growth resources by the mixture always exceeds that by the corresponding sole stands. The two components also complement each other through more effective use of growth resources by utilizing them at specific times due to difference in growth pattern and resource requirement. Indeed, there are substantial opportunities for temporal complementarities if species make their major demands on available resources at different times, thereby reducing the possibility of competition.

Supplementary interaction

If the two components interact in such a way that the yield of one component exceeds the yield corresponding to its sole crop without affecting the yield of other component, the interaction is known to be supplementary in nature. Therefore, there will be an additional yield from one of the components; thus the overall yield of agroforestry system will be more than that of sole crops of both the components. In this interaction one component is maximizing its resource utilization without limiting it for the other component. This is an independent relationship.

Competitive interaction

In this system the tree and crop components interact in such a way that the increase in the yield of one component leads to decrease in the yield of other component due to competitive interaction. Although agroforestry is envisaged as a system of plant species that benefit each other mutually or unilaterally, it is too optimistic to assume that all types of competition can be eliminated in these systems, especially in areas with poor soils and scanty rainfall. The intensity of competition is the greatest when requirements are similar, and the growth and development proceed synchronously for both the components. Thus, there is extensive overlap between species in their resource requirements, resulting in a severe competition. To reduce the competition in an agroforestry system, the tree and crop components should be selected in such a manner that they can utilize the finite resources (water, nutrients, solar energy, etc.) at different times and show different growth and development phases from each other. Competition between trees and crops is a long term problem in plantations when the crop species is perennial.

FACTORS AFFECTING INTERACTIONS

The interactive relations among the components of agroforestry are affected by many factors such as choice of species, density and age of trees, site factors, management practices, etc. Many of these factors can be manipulated for better production from the system.

Species of trees and crops

The growth pattern of trees varies from species to species and thus, affects the interactions with the associated components. Some crops may perform better in association with a particular tree component, whereas the yield of other crops may reduce with the same tree component, because different crops interact differently with the same tree species. It is not only that different crops exhibit differently with the same tree species but also the same crop may be affected differently by the different tree components.

Density of trees

Canopy cover of trees intercepts light depending upon the density of trees and consequently affects the performance of the under-grown crops. Generally, the yield of under-grown crop is decreased with increasing density of the trees. The characteristics of under-grown crop species will also determine the impact of canopy cover. Trees also modify microclimate of crops grown below and improve the physical conditions and fertility of the soil. Thus an increase in tree density should accrue some benefits while increasing the competition for other resources at the same time. Therefore, optimum density in agroforestry would vary with variation in the components.

Age of trees

The demand for various growth factors and therefore competition offered by the components would be affected by the age and the growth stage of the components. Trees take long time to attain full size and stature, and thus no or less competition is usually offered to intercrops during early years of plantation. As the trees grow, their effect on crop growth becomes apparent. Due to initial slow growth of trees, even crop or grass component may dominate and affect the tree growth adversely. For example, *Ziziphus* spp. in initial years cannot withstand competition from established pasture of *Cenchrus ciliaris*; however, once *Ziziphus* bushes attain good growth in 2 years, *Cenchrus* pasture does not affect their growth.

Site factors

Variations in climatic (rainfall, temperature, humidity, etc.), edaphic (physical, biological and chemical properties) and physiographic features of an area affect the plant growth which leads to variation in the interactive relationship of component species in an agroforestry system.

Management practices

The interactive relationship between the components in an agroforestry system also varies with management practices adopted for the system. Several characteristics could be identified as desirable attributes for trees in the agroforestry systems but often it is not possible to choose trees with all these characteristics due to various reasons. Through various management practices the interactions between the components in an agroforestry system can be manipulated. For fodder tree species, lopping done for harvesting forage periodically can at the same time increase light availability to the under-grown crops and improve their performance.

TYPES OF INTERACTIONS

Trees are the most important of all components in agroforestry systems. Based on components, interaction types may be 'tree-crop interactions' (TCI) and 'tree-animal interactions' (TAI). TCI and TAI may be positive (beneficial) or negative (harmful). These positive or negative interactions can be direct or indirect or both. Both positive and negative interactions may further be divided into above- or below-ground interaction in case of TCI.

The main effects of positive interactions in agroforestry systems are increased productivity, improved soil fertility, nutrient cycling, soil conservation, water conservation, weed control and microclimate improvements. Similarly, the effects of negative interactions include light competition, nutrient competition, water competition, pests and diseases and allelopathy. However, these interaction processes are interdependent and the manifestation of their effects is influenced by the environmental conditions. The relative importance of each effect depends on both the types of agroforestry systems and site factors.

Positive interactions

Increased productivity: The biomass productivity of an agroforestry system is generally greater than that of an annual system. The basis for the potentially higher productivity is due to the capture of more growth resources (*e.g.*, light and water) or due to improved soil fertility. The presence of trees causes shade, which in turn leads to a net effect of complex interactions resulting into microclimate amelioration, thus affecting transpiration rate, photosynthesis and energy balance of the associated crops. All these factors may translate into increased productivity from the system.

Improved soil fertility: The potential for micro-site enrichment by some trees is an extremely important aspect of agroforestry. *Faidherbia albida* in West Africa, *Pinus caribea* in the savanna and *Prosopis cineraria* in Rajasthan are known to accumulate mineral nutrients due to the process of capture of nutrient rich litter. Alley cropping using fast growing, nitrogen-fixing trees (*e.g.*, *Leucaena leucocephala* and *Gliricidia sepium*) in the humid tropics can substantially increase the soil fertility. A major feature of this system is the capacity of the trees to produce a large quantity of biomass for green manure and the need for regular pruning to prevent shading to the intercrops.

Nutrient cycling: Soil nutrients are considered among the least resilient components of sustainability. A fundamental principle of sustainability therefore is to return to the soil the nutrients removed from it through harvests, run-off, erosion, leaching and other loss pathways. In agroforestry systems, trees provide additional nutrient inputs through biological nitrogen fixation and deep nutrient capture. Deep nutrient capture is the uptake of nutrients by tree roots acting as a 'safety net' from depths where crop roots are not active. Similarly, taking up of nutrients from weathered minerals in deeper layer by tree roots is called as 'nutrient pumping' or 'nutrient mining'. These nutrients are considered as additional inputs in agroforestry systems, because such nutrients would otherwise leach down as far as the crop root zone is concerned. They become an input upon being transferred to the soil *via* tree litter (leaves, pruned materials, root debris, etc.) decomposition. The trees also improve soil health by providing huge quantity of organic matter through litter fall. Litters of high quality (low C:N ratio, low lignin and polyphenolic content) are decomposed rapidly and make nutrients available to the crop and the trees. In some cases, nitrogen is supplied by tree roots to crop roots, either due to root decay or root death following tree pruning or by direct transfer if nodulated roots are in close contact with crop roots.

Soil conservation: Contour hedgerows are highly effective in controlling soil erosion. The woody hedgerows provide a semipermeable barrier to surface movement of water, whereas litter mulch from trees reduces the impact of rain drops on the soil and minimizes splash and sheet erosion. Tree species that provide a more effective physical barrier to erosion and produce mulch material that has a longer lasting protective role should be selected for agroforestry systems.

Water conservation: The presence of trees has positive effect on water budget of the soil and crops growing between or beneath them. A mulch or litter layer increases the infiltration of rain water while simultaneously reduces evaporation from the soil. Mulch is important, especially on sandy soils as it maintains soil moisture during the dry season where water supply for the crops is a problem.

Weed control: Another potentially positive interaction in agroforestry systems is related to weeds. Shade suppresses light-demanding weeds. Weed biomass yield is positively correlated with available solar energy. Apart from shading, weed suppression is also determined by factors such as land-use history, weather and mulch quality and crop competitiveness. Shade also reduces dry-season fire risks.

Microclimate improvement: In agroforestry systems, microclimate amelioration results primarily from the use of trees for shade, live fences or windbreaks and shelterbelts. The presence of trees reduces heat and light. Atmospheric temperature, humidity, movement of air, soil temperature and soil moisture are changed due to agroforestry systems. The nature of the understorey crops determines the extent of beneficial effects of shade trees. The smaller temperature fluctuations under shade are attributed to reduced radiation load on the crops during the day and to reduced heat loss during the night. A reduction in vapour pressure deficit causes a reduction in transpiration and hence reduced water stress for intercrops. This is especially beneficial during short periods of drought. Microclimate amelioration also reduce the disease and pest pressure by facilitating biological control agents.

Tree-animal interface (positive): The positive interactions at the TAI can affect overall system productivity in various ways. Some autotrophic production that is of no direct use to the farmer such as weeds or tree fodder can be transformed into animal biomass with high nutritional and monetary value. The productivity of the tree and herbaceous crop components can be increased through the addition of animal manure as a fertilizer source. Trees provide shade to the animals in the tropics. Heat stress on the animals in the tropics significantly reduces the time spent by them in grazing in open which leads to reduction in total feed intake. Shade reduces energy expenditure of the animals for thermoregulation and this is the main reason why animals in shade generally show higher feed conversion and ultimately higher weight gain or milk production. Shade may also have a beneficial effect on animal reproduction.

Negative interactions

As all the components of an agroforestry system utilize the same reserves of growth resources, negative interactions are likely to occur in every plant association. The major yield decreasing effects at the TCI arise from competition for light, water, and nutrients, as well as from interactions *via* allelopathy.

Competition for light: In agroforestry systems, shading by the trees reduces light intensity at the crop level and this is the most limiting factor in many situations, particularly those with relatively fertile soils and adequate water availability. The relative importance of light will decrease in semiarid conditions as well as on sites with low fertility soils. Since crops differ in their responses to poor nutrition, competition for light may either be reduced or amplified by a shortage of nutrients.

Competition for nutrients: An agroforestry system is assessed mainly by the yield of crop component and its reduction receives more attention than those of the associated tree species. The effect of nutrient competition is more severe for the crop components than for trees, because the crop root system is usually confined to the soil horizons that are also available to the roots of the trees but trees can exploit soil volume beyond the reach of the crop. Therefore, the effects of nutrient competition are more severe for the crop components.

Competition for water: Competition for water is likely to occur in most agroforestry systems at some period of time except in areas with well distributed rainfall or continuous supply of ground water. The competition depends on the severity of the drought and the drought tolerance of the plants. It plays a major role in the productivity of agroforestry systems despite the use of drought-tolerant and drought-adapted plants, especially in dry areas.

Microclimatic modification for pests/diseases: Bacterial and fungal diseases may increase in shaded, more humid environments in an agroforestry system. Reduced temperature and humidity fluctuations under shade can also have a favourable effect on spread of pests and diseases. Trees and crops can be a host of each other's insect pests and diseases.

***Allelopathic interaction*:** Allelopathy refers to the inhibition of growth of one plant by chemical compounds (allelochemicals) that are released into the soil from neighbouring plants. Allelochemicals are reported to be present in practically all plant tissues, including leaves, flowers, fruits, stems, roots, rhizomes and seeds. The effects of these chemicals are dependent mainly upon the concentration as well as the combination in which one or more of these substances is released into the environment. The organic compounds released this way are often phytotoxins. Such toxic chemicals may be injurious to microbes and even to the seedlings of those plants releasing them. The effects of the chemicals may result in complete inhibition of growth or retarded growth. When the toxic exudates of the adult trees of a particular species suppress and eventually kill their seedlings, then the phenomenon is called autoallelopathy. Allelopathic compounds may be released into the environment by volatilization, leaching from living or dead tissues, exudation from roots and decay of plant tissues. The mechanism of action of allelochemicals is diverse and includes

1. Inhibition of cell division and elongation
2. Inhibition of action of gibberellins or indole acetic acid that induces growth
3. Reduction of mineral uptake
4. Retardation of photosynthesis
5. Inhibition or stimulation of respiration
6. Inhibition or stimulation of stomatal opening
7. Inhibition of protein synthesis and changes in lipid and organic acid metabolism

Agroforestry tree species having allelopathic effect on crop plants

Tree species	Annual crops whose growth is inhibited
Alnus nepalensis	Soybean
Casuarina equisetifolia	Cowpea, sorghum, sunflower
Eucalyptus tereticornis	Cowpea, sorghum, sunflower, potato
Gliricidia sepium	Maize, rice, tropical grasses
Leucaena leucocephala	Maize, rice, cowpea, sorghum, sunflower

***Tree-animal interface (Negative)*:** The most important negative interactions between animals and plants have direct effects. Low quality tree fodder with toxic compounds can adversely affect livestock production. Mechanical damage of trees or deterioration of soil properties through compaction by the animals may have a negative impact on the woody perennial component. Many tree fodder species in semiarid areas contain secondary compounds that reduce the feed value. The presence of high levels of phenolic compounds (tannins) or strong odours found in the leaves of species such as *Cassia siamea* and *Gliricidia sepium* may reduce palatability or acceptability of the fodder. Digestibility of some fodder may be low as the leaves contain toxins or toxic concentrations of certain micronutrients. Some of the harmful compounds reported from the tree fodders are mimosine in *Leucaena*, cyanogenic glucosides in *Acacia* species and robitin in *Robinia*.

MANAGEMENT OPTIONS TO NEUTRALIZE NEGATIVE INTERACTIONS

The magnitude of interactive effects between trees and other components of agroforestry systems depends on the characteristics of the species, their planting density and spatial arrangement and management of the trees. The goals of management practices should be to increase the production of the desired products and to decrease competition of undesired components.

Management options to achieve increased growth of components in agroforestry systems are microclimate amelioration, fertilization, application of mulch or manure, irrigation, soil tillage, adapted species and supplemental feeding. Similarly, management options for decreased growth include pruning, pollarding, root pruning, trenching, excessive shading, application of herbicide and grazing or browsing. Manipulating densities and arrangements is probably the most powerful method for capitalizing on beneficial effects of trees while reducing negative ones. But when trees are used as supports for crop plants, the planting density of the trees is determined by the planting density of the crops. In such conditions, knowledge of the light transmission characteristics of the tree crowns and of the options for tree management becomes important.

In many cases one cultural treatment will accomplish both the goals simultaneously, *e.g.*, pruning trees in alley cropping and applying the biomass to the soil. Though pruning will reduce the tree's competitive ability, it will also increase the growth of the associated intercrop by providing green manure and by allowing more light to penetrate to the crop.

Under conditions of severe below-ground competition, root pruning operations or trenching may eliminate or at least strongly reduce the negative effects of the trees on the intercrop. Trenching also lowers the moisture use by crops. In poorly fertile soils, addition of fertilizer enhances the productivity of components.

Fallowing the land can be beneficial where residual effects of the trees benefit the crops in subsequent years. Choice of tree species is crucial with regard to shading effects, root competition or provision of useful products for the farmer. As trees generally have a long lifetime, a good choice is a far-reaching decision, which has effects in the longer term.

Chapter 6

Selection of Tree Crop Species for Agroforestry

Farmers have been growing trees for different purposes for thousands of years. Tree species that are grown to provide more than one significant function are called multipurpose trees. These functions may be productive such as producing fuelwood, timber, fibre, fodder, food, medicine, etc. and/or protective such as soil conservation, shade, shelterbelt, microclimate amelioration, land sustainability, biodiversity preservation, etc. All trees are multipurpose; some, however, are more multipurpose than others. Tree species can be multipurpose in two ways.

1. A single tree can provide more than one function. For example, *Gliricidia sepium* is grown as living fences that provide fuel, fodder and green manure for agricultural crops - all at the same time.
2. Trees of the same species, when managed differently, can provide different functions. For example, *Leucaena leucocephala* is managed so that some trees will mainly yield wood while others mainly produce leaf fodder.

Farmers can grow multipurpose trees (MPT) in various combinations with other crops, as in agroforestry, in block plantations of trees or in naturally regenerating tree farms. In certain cases, multipurpose trees are grown and managed for only one purpose. For example, *Gliricida sepium* is grown only to provide shade in coffee plantations. The same species may be planted in some other places and is managed differently for a very different use.

BENEFITS FROM MULTIPURPOSE TREES

The tree has both productive and protective functions, and it can play an important role in enhancing the productivity of the lands to meet the demand of ever-growing human and livestock population as discussed below.

Food

1. Food for man from trees in the form of fruits, nuts, cereal substitutes, etc.
2. Feed for livestock from trees

3. Enhanced food and feed production from crops associated with trees as in various agroforestry systems through biological nitrogen fixation, better access to soil nutrients brought to the surface through deep tree roots, improved availability of nutrients due to high cation exchange capacity of the soil and higher organic matter in the soil
4. Enhanced sustainability of cropping systems through soil and water conservation by arrangements of trees to control runoff and erosion
5. Indirect production benefits through microclimate amelioration

Water
1. Improvement of soil moisture retention in rainfed croplands and pastures through improved soil structure and microclimate effect of trees
2. Regulation of stream flow for reduction of flood hazards
3. Improvement in drainage from waterlogged or saline soils by trees with high water requirements
4. Increased biomass storage of water in fodder trees for animal consumption in dry season

Energy
1. Fuelwood for direct combustion
2. Gas from wood or charcoal
3. Ethanol produced through fermentation of high-carbohydrate fruits
4. Augmentation of wind power using appropriate arrangements of trees
5. Oils, latex and other combustible saps and resins

Shelter
1. Building materials for shelter construction
2. Shade trees for people, livestock and shade loving crops
3. Windbreaks and shelterbelts for protection of settlements, croplands, pastures and roadways
4. Live fences and fence posts

Raw material for industries
1. Raw material for pulp and paper industry
2. Tannins, essential oils and medicinal ingredients
3. Wood for agricultural implements
4. Fibre for weaving
5. Wood for a variety of craft purposes
6. Fruits, nuts, etc. for food processing industries

Cash

1. Direct cash benefits from sale of tree products
2. Indirect cash benefits from increased productivity through associated crops and/or livestock

Environment

1. Production of oxygen during photosynthesis
2. Absorption of dangerous chemicals and other pollutants from the soil
3. Carbon sequestration through carbon dioxide absorption during photosynthesis
4. Reduction of air pollution through absorption of air pollutants
5. Reduction of noise pollution

SELECTION OF APPROPRIATE SPECIES

It is important to select the most suitable tree species since it is not easy to replace them once they have been planted. The success of growing MPTs mainly depends on the choice of the species in an agroforestry system. The following factors should be kept in mind while selecting tree species.

Environmental adaptation: A multipurpose tree must be able to adapt to the climate, soil, topography, and plant and animal life of the area. This is especially important for exotic species.

Needs of farmers: The selected species should meet the needs of farm families. Farmers must be involved in selecting species. The cost of acquiring planting materials should also be kept in mind.

Ease of maintenance: Some species require more attention and management than others. Farmers should consider beforehand how much time they can spare for the trees. If they require additional skills and knowledge to grow a particular species, training or demonstration programmes should be organized.

Availability of genetic materials: Seeds or seedlings of the species being considered must be easily obtained. If vegetative propagation is required, farmers should receive training on how to do this.

CHARACTERISTICS OF MPTS SUITABLE FOR AGROFORESTRY

In the agroforestry context, MPTs are those trees and shrubs which are deliberately grown and managed for more than one preferred products and/or service. The cultivation of these trees is usually economically but also sometimes ecologically motivated in a multiple output land use system. The

MPTs are the most distinctive component of agroforestry and the success of agroforestry as a viable land use option depends on exploitable potential of these multipurpose trees. MPTs for agroforestry systems should have the following characteristics.

1. They should be adaptable to local climatic conditions.
2. They should possess self-pruning properties.
3. If not self-pruning, they should be able to tolerate relatively high incidence of pruning. Their photosynthetic efficiency should not significantly decrease with heavy pruning.
4. They should be light-branching in habit.
5. They should be tolerant of side-shade in the early stages of growth.
6. Their canopy should permit the penetration of light to the ground.
7. Their phenology, particularly with respect to leaf flushing and leaf-fall, should be advantageous to the growth of the annual crop in conjunction with which they are being raised.
8. Their rate of litter fall and litter decomposition should have positive effects upon the soil.
9. Their 'above ground' changes over time in structure and morphology should be such that they retain or improve those characteristics which reduce competition for solar energy, nutrients and water.
10. Their root systems and root growth characteristics should ideally result in the exploration of soil layers that are different to those being tapped by the agricultural species.

ROLE OF MPTS IN AGROFORESTRY

Alley cropping

MPTs planted as hedgerows between rows of agricultural crops reduce soil erosion. When planted on slopes, alley crops slow down run-off rainwater and trap sediment, which can form natural terraces after several years. Farmers should prune hedgerows regularly to prevent them from competing with the crops for sunlight and water. Leaves from the pruned branches are used as animal fodder and branches are used as fuel. The tops of the hedgerows are also pruned every 6–8 weeks to obtain green leaf manure. Hedgerows may be planted in single lines or double lines depending on how much land is available and the slope of the land. Desirable characteristics of species for alley cropping are as follows.

1. Easy to establish with minimum labour for planting and maintenance
2. Fast growing so that benefits become available to the farm family as soon as possible

3. Good sprouting habit after pruning
4. Ability to fix nitrogen
5. Ability to provide palatable and acceptable fodder with high nutritive value
6. Ability to produce heavy foliage for green manure
7. Deep root system to draw nutrients and water from lower soil layers
8. Easy to propagate

Homegardens and other multistorey systems

Mixed plantings of annual agricultural and perennial tree crops around dwellings, often termed as homegardens, are a common type of multistorey agroforestry system. In homegardens, the lowest level often consists of vegetables or root crops; the second level includes fast growing trees or crops such as bananas and papaya; a third higher level may consist of large trees that provide fruit, timber and shade. Homegardens also provide a pleasant shaded living area. Many farmers grow multipurpose trees in their homegardens for flowers and fruits. Trees grown mainly for food should not be pruned regularly for fodder or fuel. Pruning can interfere with flowering and fruiting. Desirable characteristics of species for homegardens and other multistorey systems are as follows.

1. Understorey crops require the right amount of light for optimum growth. Thus the crown of trees should be high, small and open having sparse foliage.
2. They should have deep root system to draw nutrients and water from deeper soil layers so that they would not compete with shallow rooted crops.
3. Roots should not spread laterally too far from the trunk to minimize competition with nearby crops.
4. They should fix nitrogen to grow well under adverse conditions and help improve soil fertility.

Fuelwood

Fuelwood tree species should be included in agroforestry systems to reduce pressure on natural forests and save forests from denudation. Species for fuelwood should have the following characteristics.

1. Adapt quickly and yield a maximum volume of wood in a short time
2. Stabilize the soil and maintain its fertility
3. Require minimum time and effort for management
4. Tolerate diseases and pests

5. Tolerate drought and other climatic stresses
6. Ability to good coppicing
7. Have other uses that contribute to the farm enterprise
8. Produce stems without thorns with branches having a diameter small enough to be cut with hand tools and be easily transported
9. Produce wood that splits easily
10. Have low moisture content or dry relatively quickly
11. Produce minimal and nontoxic smoke (low ash and sulphur content) when burnt
12. Not spark while burning, for safety reasons
13. Produce dense wood, to burn longer

Species often used for fuelwood include *Casuarina equisetifolia*, *Acacia auriculiformis*, *Gliricidia sepium*, *Calliandra calothyrsus* and *Leucaena leucocephala*.

Fodder

Agroforestry practices that combine tree plantations with livestock are found generally in arid and semiarid regions, where natural grasslands and farm sizes are larger. In such systems, farmers allow their animals to graze on forage grasses in stands of trees. In humid and subhumid regions, which have smaller land holdings, alley cropping and other 'cut and carry' methods for procuring fodder are practised. Integrating tree growing with livestock production increases the production of meat, milk and other animal products without sacrificing agricultural land, reduces surface soil erosion, reduces reliance on inorganic fertilizer by using animal manure and provides additional income through sale of livestock. Desirable characteristics of species for fodder production are as follows.

1. Produce leaves and/or pods that animals like to eat
2. Improve fodder quality through high protein and nutrient content
3. Withstand lopping, pruning, pollarding and coppicing
4. Grow quickly, especially in early stages
5. Adapt easily to different sites and environments
6. Withstand intense shade if planted with other species
7. Free from accumulation of toxic substances in leaves
8. Fix nitrogen to enhance soil fertility
9. Withstand damage by pests, diseases and browsing animals
10. Have other uses

Some species often used for fodder include *Acacia tortillis, Faidherbia albida, Albizia lebbek, Leucaena leucocephala, Acacia nilotica, Calliandra calothyrsus, Artocarpus heterophyllus, Gliricidia sepium, Prosopis cineraria, Dalbergia sissoo* and *Sesbania grandiflora*.

Roundwood (poles and posts)

Eucalyptus spp., *Acacia auriculiformis, Acacia mangium, Albizia lebbek* and *Casuarina equisetifolia* are the important species often planted for roundwood to be used as poles and posts. Species grown for this purpose should have the following characteristics.

1. Have a straight stem
2. Have a single main stem rather than multiple stems
3. Produce few and thin branches
4. Be able for self pruning
5. Produce wood without knots
6. Have little taper from bottom to top
7. Have bark that strips easily
8. Produce wood that is durable, light and capable of supporting heavy cross loads
9. Resist termites and other wood borers
10. Be able to absorb chemical preservatives easily

Live fences

In many places, farmers plant multipurpose trees in rows along farm boundaries as 'living fences'. In addition to providing fodder and fuelwood, living fences provide privacy and protection from browsing animals. Species grown for this purpose should have the following characteristics.

1. Grow well under adverse conditions
2. Grow well in close spacing
3. Require minimal maintenance
4. Withstand lopping and trimming
5. Ensures fast establishment through vegetative propagation while reducing the chance of spreading to cultivated areas
6. Grow fast to a medium height
7. Be long lived
8. Have prickles or stiff branches and leaves that animals do not like to eat
9. Should not have adverse effects on other tree species or crops they are associated with

Species often used for live fences are *Cajanus cajan, Erythrina poeppigiana, Calliandra calothyrsus, Gliricidia sepium, Casuarina equisetifolia* and *Pithecellobium dulce*.

Windbreaks and shelterbelts

Windbreaks are strips of trees planted closely together along the edges of croplands perpendicular to prevailing winds. Especially in dry areas, windbreaks can provide protection to crops and soils from the detrimental effects of wind. Species grown for windbreaks and shelterbelts should have the following characteristics.

1. Tolerate harsh environments
2. Withstand strong winds
3. Have a bushy, deep crown that still allows some wind penetration
4. Keep lower limbs for a long time
5. Have strong roots
6. Grow quickly
7. Live long
8. Tolerate pests and diseases
9. Should not have roots that compete excessively with nearby crops for water and nutrients

Species often used for windbreaks and shelterbelts are *Acacia nilotica, Casuarina equisetifolia, Azadirachta indica* and *Erythrina peoppigiana*.

Soil protection and rehabilitation

The important species often planted for soil protection and rehabilitation are *Acacia auriculiformis, Casuarina equisetifolia* and *Prosopis juliflora*. Species for this purpose should have the following characteristics.

1. Have faster growth rate
2. Grow well under adverse conditions
3. Have a spreading crown
4. Have a vigorous root system to bind the soil and a strong taproot especially in the areas prone to landslides
5. Associate well with nitrogen fixing organisms and mycorrhizal fungi in the soil
6. Reproduce vigorously, for example, from root suckers or through abundant natural seed fall and seedling development
7. Tolerate fire

Industrial uses

Species grown for sawn timber, plywood, veneer, pulpwood or other industrial uses should have the following characteristics.

1. Have faster growth rate
2. Have a maximum rate of growth in early stages of development
3. Have a straight stem, uniform size and small branches
4. Show good natural pruning with rapid healing of wounds if used for producing sawn timber and plywood
5. Have a good physical, mechanical seasoning, preserving and processing properties if used for producing sawn timber
6. Have wood that is easy to peel, slice and free of knots if used for producing plywood and veneer
7. Have wood with a high density and long fibres if used for producing pulp

The species suitable for industrial uses are *Pinus* spp., *Tectona grandis*, *Populus* spp. and *Terminalia* spp. for sawn timber; *Dipterocarpus* spp. and *Tectona grandis* for plywood and veneer; *Acacia* spp, *Leucaena* spp, *Pinus* spp. and *Eucalyptus* spp. for pulpwood.

Shade and nurse trees

Species often used as shade or nurse trees are *Erythrina poeppigiana*, *Gliricidia sepium*, *Leucaena leucocephala* and *Sesbania grandiflora*. Species grown for this purposes should have the following characteristics.

1. Have fast early growth
2. Establish easily, preferably by cuttings
3. Be evergreen
4. Live long
5. Tolerate soil compaction by animals walking and sitting beneath them, as happens on pasture land
6. Coppice well and withstand some lopping
7. Fix nitrogen in the soil to enhance soil fertility
8. Have a dense and spreading crown, if used to shade grazing animals
9. Have a light crown, if used as nurse trees for other crops
10. Should not compete with crops for soil nutrients if used as nurse trees

Nitrogen fixing trees

Nitrogen fixing trees (NFTs) are valuable in subtropical and tropical agroforestry. They can be integrated into an agroforestry system to restore nutrient cycling and fertility self-reliance. NFTs are often deep rooted, which allows them to draw nutrients from subsoil layers. Constant leaf drop from them nourishes soil life, which in turn, can support more plant life. The extensive root system stabilizes soil and adds organic matter to the soil. There are many species of NFTs that can also provide numerous useful products and functions, including food, wind protection, shade, animal fodder, fuelwood, living fence and timber, in addition to providing nitrogen to the system. Many fast growing NFTs can be cut back regularly for mulch production. The NFTs may be integrated into an agroforestry system in many different ways including clump plantings, alley cropping, contour hedgerows or shelterbelts. They can serve many protective functions such as microclimate amelioration for shade loving crops like coffee; trellis for vine crops like vanilla, pepper and yam; and living fence. The major nitrogen fixing tree genera adopted in agroforestry systems are *Acacia, Albizia, Alnus, Casuarina, Dalbergia, Erithrina, Gliricidia, Leucaena, Prosopis, Sesbania*, etc. NFTs should have the following characteristics.

1. Ability to fix nitrogen at higher rates
2. Have faster growth rate
3. Ability for coppicing
4. Ability to survive in edaphic and climatic stress situations
5. Produce more seeds and easy to propagate

Other uses of MPTs

Trees can help reduce soil erosion along streams and gullies. Their roots serve to hold the soil in place and reduce the impact of storm water. Some tree species suitable for stabilizing stream banks and gullies are *Paraserianthes falcataria, Gmelina arborea, Leucaena leucocephala, Sesbania grandiflora* and *Moringa oleifera.*

Some multipurpose tree species can also be used for decoration. These should have attractive flowers and fruits that can be eaten by birds and small animals. The species include *Acacia auriculiformis, Casuarina equisetifolia, Pterocarpus indicus, Albizia saman* and *Tamarindus indica.*

Trees can also be planted to produce other commodities such as *Eucalyptus* spp. for essential oils and honey, *Acacia* spp. for tannin for making dyes, *Azadirachta indica* for bio-pesticides and *Hevea brasiliensis* for latex.

SUITABLE MPTs FOR AGROFORESTRY UNDER DIFFERENT CLIMATIC CONDITIONS

Hot desert: *Acacia tortilis, Capparis* spp, *Prosopis chinensis, Prosopis cineraria, Tecomella undulate*

Cold desert: *Populus alba, Populus cilita, Populus euphretica, Populus nigra, Populus tremula, Salix alba, Salix fragilis*

Tropical semi-arid: *Acacia nilotica, Acacia senegal, Acacia tortilis, Albizia lebbeck, Azadirachta indica, Eucalyptus camaldulensis, Prosopis* spp, *Salvadora persica, Tamarix* spp

Subtropical semi-arid: *Acacia modesta, Albizia procera, Bauhinia variegata, Ficus* spp, *Morus indica, Pinus roxburghii*

Temperate semi-arid: *Corylus colurna, Juniperus macropoda, Pinus gerardiana*

Humid tropical: *Areca catechu, Artocarpus heterophyllus, Chukrasia tubularis, Cocus nucifera, Gmelina arborea, Pterocarpus santalinus, Schima wallichii, Tectona grandis, Terminalia alata, Terminalia myriocarpa*

Humid subtropical: *Acer oblungum, Acrocarpus fraxinifolius, Aesculus indica, Eucalyptus globulus, Prunus* spp, *Quercus* spp

Humid temperate: *Acer campbelii, Alnus nitida, Pinus alata, Pinus wallichiana, Populus ciliata, Robinia pseudacacia*

Tropical sub-humid: *Adina cardifolia, Bombax ceiba, Casuarina equisetifolia, Dalbergia latifolia, Dalbergia sissoo, Eucalyptus citridora, Eucalyptus teriticornis, Leucaena leucocephala, Moringa oleifera, Morus alba, Populus deltoides*

Subtropical semi-humid: *Albizia chinensis, Celtis australis, Eucalyptus globules, Eucalyptus grandis, Grewia optiva, Morus indica, Pinus roxburghii, Toona ciliata*

Temperate semi-humid: *Acacia mearnsii, Cedrus deodara, Celtis australis, Juglans regia, Quercus* spp

SUITABLE MPTs FOR AGROFORESTRY UNDER DIFFERENT SOIL TYPES

Desert soil: *Acacia senegal, Acacia tortilis, Acacia. nilotica, Prosopis chilensis, Prosopis cineraria*

Recent alluvium: *Acacia catechu, Bombax ceiba, Dalbergia sissoo*

Saline-alkali soils: *Acacia nilotica, Ailanthus* spp, *Azadirachta indica, Eucalyptus* spp, *Pongamia pinnata, Prosopis* spp, *Tamarix* spp

Coastal and deltaic alluvium: *Areca catechu, Casuarina equisetifolia, Cocus nucifera*

Red soils: *Acacia nilotica, Azadirachta indica, Dalbergia sissoo, Dendrocalamus strictus, Eucalyptus hybrid, Leucaena leucocephala, Madhuca indica, Mangifera indica, Pterocarpus marsupium, Tectona grandis*

Black cotton soils: *Acacia nilotica, Adina cardifolia, Aegle marmelos, Bauhinia* spp, *Dalbergia latifolia, Hardwickia binata, Tamarandius indica, Tectona grandis*

Laterite and lateric soils: *Acacia auriculiformis, Azadirachta indica, Emblica officinalis, Eucalyptus* spp, *Tamarindus indica, Tectona grndis*

Peaty and organic soil: *Bischofia javanica, Ficus glomerata, Glircidia sepium, Syzygium cumini*

Chapter 7

Agroforestry Planning and Management

Agroforestry is a land use system that involves two or more plant species at least one of which must be a woody perennial. When perennial woody and herbaceous components are grown together on the same piece of land their performance would largely depend on their ability to share various growth resources in a given environmental situation. Due to difference in growth pattern and resource requirement of the components in agroforestry situation, a close interactive relation is obvious. Thus, careful management practices for the components are required to establish a successful agroforestry system. The characteristics of the trees and crops, and their interactions, can be modified with good management practices in order to take advantage of the positive characteristics and minimize the effects of the negative ones. Effective and efficient agroforestry management may be divided into two groups - tree management and agricultural crop management.

FACTORS AFFECTING THE SELECTION OF TREE SPECIES

Trees in an agroforestry system contribute significantly to the total output of the system and therefore, success of the system depends on the selection of the tree species. The following factors are to be considered to choose the best tree species for the agroforestry systems in a particular site.

Knowledge of the species

The grower should have a sound knowledge of the species he intended to include in the system. Previous successful experiences and existing preference for the species by other farmers provide a better assurance of success and adoption of the species. Trees should be adaptable to local climatic conditions.

Commercial value or local use of the species

The trees should produce timber of medium to high quality. The optimum would be to introduce timber trees with the highest quality and market prices, when the environmental and service factors and growth rate are appropriate. The marketing options and the local and external demand for the products are the most influential factors in the species selection. Some of the species provide secondary products, such as poles and firewood during pruning and thinning.

Rapid growth

Species should be fast growing so that benefits become available to the farm family as soon as possible. Rapidly growing tree species quickly generate a product and have a lower maintenance cost.

Self-pruning in open light conditions

Trees in agroforestry systems tend to have poor trunk form compared to trees growing in block plantations due to the wide tree spacing and lack of lateral competition. Trees in open conditions tend to have more branches and more persistent branches, higher harvesting costs, more waste at the sawmill and lower timber quality due to knots in the wood. However, in some timber species the lower branches are self pruned, which means they die and fall when in open conditions. These species are preferable for agroforestry systems because they have lower production costs and higher timber value.

Availability of certified germplasm

The probability of success is increased if certified seeds or seedlings are used. It is better to use high quality genetic material that produces strong and healthy plants, and the use of genetic material of dubious quality should always be avoided. Choosing healthy and vigorous plants in the nursery or during thinning of natural regeneration is also important.

Lack of susceptibility to pests and diseases

Species susceptible to the pests and diseases of the area, or with other known problems, must be avoided. The trees can host pests and diseases that damage the agricultural crops.

Minimal negative effects for the associated crops

Even with crops that are generally grown under shade there are sometimes negative effects from competition for light, nutrients and water. The leaves of some trees can contain substances toxic to some crops. These types of problems are almost inevitable in permanent agroforestry systems. In addition to managing the competition it is important to make sure that the trees do not have alleolopathic effects on the associated crops.

Small and open crown

In order to minimize the competition for light with associated crops the tree species should have open and narrow crown. Harvesting of the trees is less expensive and there is less risk of damage to the crops since the majority of damage is caused by the fall of the crown rather than the fall of the trunk. It is preferable to have non-continuous foliage that provides patches of shade rather than a uniform canopy that creates a low quality light environment for the crops. It is an advantage if the species has small leaves in order to avoid the effect of unifying the raindrops into strong steady streams of water that fall off the leaves, causing damage to both crops and soil.

Deep rather than surface root system

A root system on the soil surface creates strong competition with crops and is more susceptible to damage during crop management. The root systems and root growth characteristics should ideally result in the exploration of soil layers that are different to those being tapped by the agricultural species.

TREE PROPAGATION METHODS

Trees that provide materials for propagation are called mother trees. Selection of mother trees is important because the young trees will be expected to inherit the favourable characteristics from the mother tree, such as fast growth, upright or spreading shape of the tree crown, good flowering and fruiting, and tolerance of diseases or pests. There may be sexual propagation by seed or vegetative propagation by using other parts of the tree. The genetic make-up of the seeds differs due to recombination of the genes, resulting in seedling variation. Thus, all the seedlings may look like the mother tree in some respects, but no two seedlings are the same. In case of vegetative propagation through cutting, layering, grafting or budding, the genetic make-up of the young tree is exactly the same as that of the mother tree. Consequently, all new trees grown vegetatively from one mother tree have the same set of gene pairs and therefore the same characteristics; they form a so-called clone. Differences between plants of a clone can only be caused by different growing conditions.

The number of trees can be increased through natural regeneration of existing trees or through man's propagation activities. Spontaneous growth of new plants, without the interference of man, is called natural regeneration. Seeds are the main source of new plants in natural vegetation. Some plants also spread through suckering. Regrowth from the stump after cutting down certain trees (*e.g.*, many *Eucalyptus* species) is also a way of natural regeneration. Natural regeneration can be encouraged by improving conditions for growth of young trees and by protecting them against hazards such as fire and browsing animals.

Farmers may collect the seed and sow it where they want these trees or shrubs to grow. Instead of collecting and sowing seed, young seedlings or rooted suckers may also be gathered from where they are growing spontaneously for direct field planting. Field establishment of these wildings needs care and favourable growing conditions, because these plants were not raised to be transplanted and often most roots are lost when the plants are lifted.

If natural regeneration or direct field planting is not successful for the tree species, the trees must be multiplied in a nursery. A nursery is a plot of land where young trees can be raised under more or less controlled conditions like protection against animals and birds, an assured irrigation facility, shade against the hot sun, and shelter against strong wind, and improved soil conditions. It takes about from two months to a year to raise seedlings of different agroforestry species in the nursery, depending on how fast the plants grow and on the preferred size for main field planting.

Seed

Seed can be collected locally or obtained from distributors, seed centres or research institutes. Success of planting mainly depends on seed quality. The seed should be available at the right time in sufficient quantity. If collection is local, it is important to know where the best trees are located. Seed should come from an area that is as similar as possible in altitude, quantity of rainfall, type of soil, etc. to the area where the seedlings will be grown. Seed should be collected from healthy and vigorous mother trees with the desired characteristics. Seed should not be collected from isolated bushy or very young trees. Seeds should be collected only from ripe fruits. If fruits are collected too early, the seed might be immature. If collection is delayed too long, the seed may be attacked by insects and fungi. Seeds of some tree species lose their viability if they are stored even for a week after their extraction from the fruit. These seeds are called recalcitrant seeds. Thus, recalcitrant seeds should be left in the fruit till they are sown. Once the seeds have been extracted from the fruits they are to be cleaned thoroughly to remove all bits of fruit flesh, pulp or husk. Seeds are sieved to remove undersized and broken seeds. In most cases, when seeds are immersed in water the good ones sink and bad ones will float. Then the good seeds are removed and dried in shade carefully.

If seeds cannot be planted immediately after collection and extraction, they should be stored carefully. Some seeds can be stored at normal temperature for many years if they are kept dry. However, recalcitrant seeds should be sown immediately after extraction from the fruit. Seeds are dried before storage to prevent infection by fungi and bacteria. But extreme heat must be avoided as this may destroy the seed viability. Seeds should not be exposed to direct sunlight. Seeds are dried in a shaded and well ventilated place.

Seed treatments to accelerate germination

Germination of seed of some tree species may take many months. Germination may also be delayed by a hard seed coat or by presence of certain substances in the seed. Such seeds may be treated to speed up germination and it is very important to ensure more simultaneous germination.

Soaking in water or acid: This is a simple method involving soaking the seeds for 2 days in water prior to sowing. This makes the seed coat soft and the substances that delay germination are leached out. In some cases acid may be added to make the liquid more abrasive. This method works for seeds from many tree species.

Hot water treatment: In this case, seeds are immersed in boiled water immediately after taking it off the fire. The water with seeds is allowed to cool overnight and seeds are rinsed with clean water the next day. This makes the hard outer cover soft and the sprout can break through it more easily.

Scarification: To facilitate germination a shallow incision is made in the hard seed coat using a file or the tip of the seed is cut off. A simple way of scarifying legume seeds is to rub them over a rough surface, *e.g.*, sandpaper. This scratches the seed coat, but the scratches should not go right through the coat.

Stratification: When trees from the temperate zones are to be grown in tropical highlands seeds of these trees may require winter chilling to break seed dormancy. Here, the seeds in a pot with moist sand are kept in the refrigerator for about two months.

Wildings

Wildings are planting materials such as seedlings, cuttings, etc. collected from natural vegetation. Wildings should have at least 2–4 fully developed leaves. Larger wildings have stems of about the diameter of a pencil. When the soil is moistened with rain water, the wildings can be collected either with or without a root ball. The network of roots and the soil clinging to them when a plant is lifted from the soil is called its root ball. Wildings with root ball are dug out by making cut on soil at both sides of the young plant at an angle of 45 degree using a spade. The plant is then lifted on the spade, holding the stem with the other hand and removed with the root ball intact. It is difficult to dig up small wildings with a root ball, because the root system is too small to hold the soil. Bare-rooted wildings are uprooted by loosening the soil around the roots with a pointed stick. The taproot and large sized roots are often trimmed. Leaves except for the top ones can be removed to reduce transpiration. The wildings should be planted in as quickly as possible after their removal.

Cuttings

The stem, leaf or root of a tree may be cut for rooting. But stem cuttings are most common. The hardwood cuttings are generally taken during the dry or cool season, when shoot growth is minimal. In deciduous trees the cuttings are best taken when the trees are leafless. The leaves are stripped from the branch or stem cutting that is inserted in the soil. Usually a few leaves are left at the tip of the cutting. The cuttings are likely to dry out if the leaf area is large. The diameter of cuttings ranges from pencil-thick to about 3 cm. The upper cut is oblique so rainwater runs off. The lower cut is usually made just below a node, because rooting generally occurs mainly at the node. There may be rotting which leads to failure of the cutting if a cut is not smooth and clean. Improper and unclean cutting may also infect the wound on the mother tree. Straight branches and twigs are preferred for cuttings; they grow upright after rooting, forming a tree with a proper trunk. Several pruning techniques such as cutting back, pollarding, and coppicing may be practised to stimulate trees to form additional shoots from which cuttings can be taken.

Cutting back: Cutting back branches or twigs stimulates buds to break and grow into new shoots. These shoots are of uniform in age, shape and size, and can be used as cuttings.

Pollarding: Pollarding is a drastic pruning where the tree trunk is cut at a height of about 2 m. A number of watershoots emerge below the cut. After about a year these shoots can be used as cuttings which are termed as live stakes. *Gliricidia sepium* or *Erythrina* species produce good live stakes. The cuttings from these species strike root easily when planted during the wet season.

Coppicing: Coppicing is cutting the tree trunk off close to the ground in order to stimulate the growth of new shoots on the stump. These shoots can then be cut off and used for planting.

Cuttings for propagation are also available from tree roots, *e.g.*, *Casuarina* species. Some trees even produce root suckers spontaneously. These suckers if detached from the mother tree and planted in another place they develop their own roots and grow to a new tree. In some tree species, roots are cut using a sharp spade to encourage the emergence of root suckers. These soot sucker cuttings are planted in a nursery under shade. If a high humidity is maintained rooted suckers will be ready after a couple of months for planting.

Layering

Some tree species cannot be propagated through stem cuttings because cuttings from these trees do not form roots readily. So in some cases, trees may be induced to form roots on their shoots before these are detached from the mother tree to be planted. This method of propagation is called layering. These techniques are particularly suited to home gardeners. In simple layering, long flexible twigs of certain shrubs are bent down to ground level and part of the twig behind the tip is covered with soil. The twig may be wounded by cutting a notch in it or by removing a ring of bark. Rooting will take place near the wound on the side of the tip. But it is difficult to bend an upright branch down to the ground to be layered. The alternative is to bring the soil up to the branch called air layering. A ring of bark is removed from the twig or branch that is to be air-layered and the soft cambium layer is scratched off, so that the wound cannot heal. A ball of moist friable soil, shredded coconut fibre or other rooting medium is fixed around the ring by wrapping it with polythene sheet so that it does not dry out. Roots are formed just above the ring and after 2–6 months the rooted air layers can be cut. In homegardens fruit trees such as guava and litchi are often propagated by air layering.

NURSERY MANAGEMENT

In an agroforestry system agricultural crops are generally directly sown in the field but planting materials of tree crops are usually raised in nurseries and the prepared seedlings are then planted in the main field. However, before going ahead with nursery plans, a thorough economic analysis is required since a good nursery can be an expensive undertaking. For small and medium scale plantations, procurement of seedlings from an authenticated commercial seedling grower is the best and the cheapest source of quality planting material.

A successful nursery operation depends on many factors like selection of site, planning and supervision, timely operations, etc. The selection of nursery site is important to ensure efficient production of good quality seedlings and easy nursery management. The following factors should be considered while selecting a site for nursery.

Water: A nursery should be established near a regular source of water. The nursery requires between 10 and 20 litres of water per square metre bed per day, depending on the temperature and the size of the young plants.

Soil: A good source of sandy loam or loamy soil, preferably forest topsoil, should be as close as possible to the site to reduce transport costs. This is particularly important if a large number of potted seedlings are to be produced.

Topography: The site should be relatively flat with 1–2% slope. The gradual slope allows runoff and prevents water logging and erosion. In hilly areas, nurseries must be built on terraces.

Microclimate: The site should not be exposed to dry winds. The site should receive sunlight for the major part of the day.

Materials: There should be easy access to manure, chemical fertilizers, compost, polybags, pesticides, etc.

Location: To minimize transportation time and cost, the nursery should be located as close as possible to the planting area.

Access: The site should be close to a good road so that the nursery will be accessible throughout the year.

Labour: There should be enough local labour available near the nursery site.

Size: This is determined mainly by the number of plants to be produced, the time they will spend in the nursery and their density in the beds.

Fence: The nursery must be protected against intruders, particularly stray animals and birds.

Shade: A nursery must be shaded. Shade prevents overheating during the hot season and raises the humidity in the beds by impeding air movement.

Raising bare-rooted seedlings and wildings in the nursery

Growing bare-rooted seedlings is the simplest form of tree propagation in a nursery. The seeds are sown in a well prepared bed and the seedlings are allowed to grow till they are ready for field planting. A more common method is to sow the seed closely in seedbeds and to prick out the young seedlings into plantbeds at the right spacing for their further growth in the nursery. Pricking out is the term used for transplanting young delicate seedlings by lifting them gently, using a pointed bamboo stick. There are good reasons to separate the seedlings at germination stage (seedbed) and to allow further growth in the nursery (plantbed).

1. The seeds of many trees germinate slowly and unevenly. By pricking out the seedlings that have reached the right stage every few days, a uniform stand is achieved in the plantbed. If plants grow uniformly they can be given the right treatment (*e.g.*, topdressing with fertilizer, root pruning) at the right time.
2. The requirements for germination are not the same as those for seedling growth. A seedbed needs careful levelling to prevent washing down of seeds to a lower edge during watering. The seedbed requires a fine tilth so that the seeds can be sown at the proper depth. The soil in a seedbed does not need to be fertile because the seedlings are pricked out before the food reserves in the seed have been exhausted. A plantbed

does not require such a perfect surface, but friable, rich soil is required to encourage growth and branching of the roots.

Seed-trays of 40 cm × 30 cm size or so and about 5 cm deep, filled with coarse sand may also be used as seedbeds. The soil in seedbeds must be compacted to ensure good contact between the top layer in which the seeds are sown and the deeper layers. Light watering after sowing helps to settle the soil around the seed. Seeds should not be sown too densely, as crowding may result in weak seedlings. If seeds are sown in seedbeds or trays and are to be pricked out later, it is better to broadcast or dibble in rows. Small seeds (*e.g., Eucalyptus* species, *Acacia mangium*) are mixed with fine sand in a sand-seed ratio of 2:1. This makes even sowing at the right density much easier. Broadcasting is limited only to small seeds, because large seeds cannot be placed at the right depth by broadcasting. Sowing in rows ensures an even distribution of seeds and facilitates weeding and pricking out.

If seeds are sown directly in their final position in the nursery without the practice of pricking out they are sown in pits at the proper spacing in a regular pattern. The spacing depends on the tree species and the required seedling size for planting out. Seed in seed-trays is often covered with coarse sand, which is easier for the seedlings to break through. If germination takes several weeks it is better to mulch seedbeds or seed-trays. The mulch softens the impact of watering and keeps the topsoil moist. However, the mulch is removed as soon as the first signs of germination are observed. After emergence of the seedlings they are thinned leaving only a single seedling in each grid position.

Pricking out

Seedlings in seedbeds or seed-trays are to be transferred to the plantbeds. Pricking out is usually done when the seedlings have one or two normal leaves. If seeds are sown at the right spacing, the seedlings fill the bed or tray by the time they have reached the stage for pricking out. For most species pricking out is done within a few weeks from emergence of the sprout. Simultaneous germination makes it possible to prick out all the seedlings at the same time. But even with proper pre-germination seed treatments, seeds of many tree species do not germinate simultaneously. Thus, it is often necessary to prick out repeatedly at an interval of several days.

In most tree species seedlings are large enough for pricking out at about 10–20 days after germination, but conifers are often pricked out sooner, even 24 days after emergence. If the seedlings are too young they will be very delicate and likely to be damaged. However, if seedlings are allowed in the seedbed for long they become overcrowded and separating the roots becomes difficult. Pricking out must be completed within a few hours per day, avoiding work during the hot hours. Successive sowings are planned in such a manner that batches reach the pricking-out stage one after the other, not simultaneously.

Raising seedlings in pots in the nursery

Sowing or planting in pots is an alternative to sowing or planting in beds. This requires extra work in the nursery and greatly increases the cost of transport if the seedlings are to be planted in a remote site. The extra expense involved in using polythene pots is compensated by better quality of the planting material, resulting in a high percentage of survival and a uniform stand in the field. The main advantage is of course that the roots are left undisturbed in the potting soil when pot seedlings are planted out.

Soil for filling the pots should be light and loose textured for good aeration and drainage, so that root development is not impeded. A commonly used mixture is 3 parts topsoil, 1 part compost and 1 part sand. The pots are watered and left to stand for a few days till the pot soil has settled. The pots are placed close together in blocks of similar length and width as the plantbeds in the nursery. The pots should always be placed upright even though it is often easier to lean them against each other. Pots can be seeded directly, or they can be planted with seedlings pricked out from seedbeds or seed-trays. Wildings or cuttings can also be planted in pots. For most tree species pricking out seedlings into pots has mainly two advantages over direct seeding in the pots.

1. Pricking out stimulates branching of the root system, so that the pots are better filled with roots to hold the soil during field planting.
2. Each batch of pricked out seedlings are uniform in height and growth.

Care for plants in the nursery

Watering: Light watering should be provided to the seedbeds regularly until the seedlings emerge. Alternatively, the seedbeds may be mulched during this period to save water and to keep the topsoil moist. Mulch is removed as soon as the first seedlings emerge.

Shade: When seeds have been sown, seedlings have been pricked out or cuttings or wildings have been planted, shading is necessary to prevent drying out and scorching of the leaves. A 50 cm-high shade frame can be erected by using bamboo or wooden sticks. Trees should never be planted within a nursery, because full light is required for hardening seedlings before planting.

Weeding: Seedbeds must be kept weed free so that the seedling roots do not get tangled up with the roots of the weeds. The competition between weed and young seedlings for nutrients and water is also checked.

Windbreak: A windbreak consisting of at least two rows of shrubs and one row of fast growing trees should be planted in areas having high wind velocity. These should be placed on the side of the nursery facing the wind. The trees should be far enough from the beds so that their crowns will not shade the beds.

Fertilizer application: Liquid fertilizer is ideal for nursery application. Chemical fertilizer is usually not recommended. The mixture of one part of cow dung and five parts of water is left for 10 days. Before spreading the liquid it should be diluted by adding more water. Seedbeds require no fertilizer, but plantbeds should receive liquid manure once in two weeks. If the soil in the plantbed is deficient in plant nutrients chemical fertilizers may be applied in small quantity.

Damping off control: Damping off is a common and serious disease in seedbeds. Several moulds in the soil may infect germinating seeds and emerging seedlings. Sowing in clean soil, such as fresh river sand, is the best way to prevent damping off. High humidity, heavy wet soil, too much shade, crowding of the plants in the seedbed, and a high organic matter content of the soil can all encourage fungus growth in a seedbed or seed-tray. The risk of damping off is reduced if the soil is well drained and aerated.

Field planting

Hardening-off: Five to six weeks before field planting, watering to the nursery should gradually be cut down and eventually stopped. Reduction of the shade can start much earlier depending on the tree species. This is known as hardening-off and is done to accustom the young plants to hardy conditions in the main field. The seedlings in the nursery may be watered on the day before planting out to facilitate lifting of the plants.

Lifting the plants: Bare-rooted plants are usually dug up by carefully lifting the plants with a fork. The plants are shaken gently to remove excess soil from the roots and packed in sacks, polythene bags with holes, banana leaves, or crates for transportation to the planting site. Diseased and plants that are not true to type should be removed.

Planting out: The best period for planting is after the first shower of monsoon rains. The land should be weed free and the soil loosened to improve the uptake of water. Planting spacing of trees depends on species, environmental conditions and the purpose for which these are grown.

Fertilizing the seedlings: The young trees are provided with extra nutrients by mixing compost, manure or fertilizer in the topsoil before or during planting. Fertilizer should be used in small quantities and only in moist soil. Manure and compost release nutrients slowly over a longer period of time and they also help to retain soil moisture.

TREE MANAGEMENT

A silvicultural regime or management strategy is a set of prescribed methods of manipulating the growth of trees so as to satisfy the objectives of the grower. These strategies must be developed with clear objectives appropriate to the existing physical and biological conditions, resources and technologies available. The main objective of tree management in an agroforestry system is to maximize the value of tree production without reducing the agricultural production significantly. The trees are to be managed in such a way that the agricultural crops do not suffer from inadequate levels of solar radiation, plant nutrients and soil moisture due to competition from the tree component. Agricultural production is expected throughout the life of the trees in an agroforestry system.

Aftercare in the main field

In an agroforestry system planting materials or seedlings of tree crops are generally raised in nurseries and the prepared seedlings are then planted in the field. Seedlings may be procured from a known source if the requirement is small. Raising a good nursery may sometimes be expensive. Thus, commercial seedlings are the best and the cheapest source of planting materials. But these must be genuine and true-to-type. Once seedlings are planted in the main field the work is certainly not finished. Care of the young trees in the field will improve establishment and growth.

Protection practices

Fire: Fire is often a serious threat during dry seasons. Regular weeding and removing dead branches and dry leaves decreases the risk of fire. This, however, affects the layer of litter, and ultimately soil fertility declines. Firebreaks (strips of cleared land which stop a fire) can be constructed between or around tree stands.

Wind: Young plants can be protected against the wind by using stakes. If the tree is moving to and fro in the wind, young roots are ruptured and this may greatly delay establishment of the plantation. The stake is placed on the windward side, so that the tree is not rubbing against the stake when the wind blows.

Shade: Many fruit and tree crops benefit from shade during the first year after field planting. If the young trees are staked, a palm frond can be tied to the stake to shade the tree.

Animals: Animals browsing leaves or twigs cause much damage to the young trees. Small animals such as rats can be caught in traps. Other solutions include digging ditches or placing wire fences. Young trees can be protected from birds by fine netting. Domestic animals, such as sheep, goats and cattle, not only browse trees, they also trample small trees down. The livestock should be kept away, if necessary by employing a guard.

Pests and diseases: Healthy plants are often able to outgrow pests and diseases. Using hardy species, healthy planting material and good maintenance techniques reduce the risk of damage. Chloropyriphos should be applied in the pits at the time of planting if white ants are a problem. Mulch containing insect repellent material such as *Azadirachta indica* or *Eucalyptus* leaves can sometimes prevent the seedlings. If insect pests or diseases are noticed immediate control measures are to be taken.

Maintenance practices

Watering: Seedlings are planted usually at the onset of monsoon. But the problem arises if monsoon ceases early or a dry spell continues immediately after the plantation. Watering should be done as and when required. The weeds around the trees should be cleaned before watering. A small basin around each tree or a trench running along a row of trees will help the water to flow towards the tree. Water should be given in the late afternoon. The quantity of water needed depends on the weather and the amount of rainfall.

Weeding: Climbing or winding weeds can quickly strangle young trees. A circle with a diameter of at least 50 cm around the tree should be kept free of weeds. This prevents competition between trees and weeds for water, light and nutrients. Apart from weeding, weed growth can also be suppressed by mulching using organic material or small stones around the tree.

Nutrition: Trees are usually fertilized twice a year, once just before the onset of monsoon and another immediately after the cessation of monsoon. Nitrogen, phosphorus and potassium are given 50 gram each per plant per year for three to four years after plantation. The fertilizer is spread in a circle around the tree and mixed with soil by hoeing superficially, taking care not to damage the roots. The nutrients applied to the agricultural crops provide the required nutrition to the tree crops once the seedlings are established. Mulching with organic material around the tree will also provide extra nutrients.

Replanting: Replacement with the seedlings of about the same size and same age is done if some of the young trees die. This is known as filling in or gapping up. Some seedlings in the nursery are retained for this purpose in seedling beds or large pots. In some agroforestry systems, such as shade trees for perennial crops like tea and coffee, or windbreaks, replanting is much more important than in pure plantations because uniform cover is fundamental to the function of the planting. In windbreaks it is particularly important to have 100% of the positions planted during the first year since open places in the line create turbulence and reduce the overall effectiveness of the windbreak. Replanting should take place within one or two months of the initial planting, especially in areas with a significant dry season, as it is difficult for trees planted in the following years to succeed due to competition with the more advanced original seedlings.

Planting spacing of trees

In case of pure timber plantation, seedlings are planted at high density with the intention of thinning the plantation later. But in agroforestry systems, trees are generally planted at their final density. In agroforestry systems it is therefore necessary to have high survival rates, rapid initial growth, and in the case of timber trees, good trunk form in all of the planted individuals. However, it may be recommended to plant more trees than required so that some degree of selection can eliminate unsuited, damaged, insect infested or diseased trees when they become evident. The genetic quality of the tree species influences the number of trees to be planted per unit area to guarantee a quality tree crop.

The initial spacing of the trees depends to a large extent on the objectives of the plantation, the growth habit of the species, site quality and management. If the farmer wants to produce firewood the initial spacing may be 2 m or less in small plantations. For saw-wood plantations the spacing should be 3 m, with later thinning. In boundary plantings initial spacing for timber trees should be 3–5 m, whereas shade trees for coffee or cacao should be at more than 8 m spacing. In agrisilvicultural systems, hedgerow spacing is maintained at 4–8 m and tree to tree spacing within the row should be 2–3 m. Windbreaks should be planted in a line perpendicular to the predominant winds, the lines separated by about 20 times the maximum height of the trees. The final tree population depends upon several factors which are to be considered before deciding final crop stock.

The relative value of tree crops and agricultural crops greatly influence the choice of stocking rate because of the competitive nature of the various components in an agroforestry system due to interactions among them. The more valuable the agricultural crop, the fewer will be the number of trees.

Performance of trees at various densities is an important consideration. A tree species performs differently when grown at different densities. At the same time trees of different species do not perform similarly when planted at the same density. High plant density leads to lower growth rate in stem diameter on individual trees. However, the total biomass production per unit area remains more or less the same over a range of stockings. The choice is either lots of more number of small diameter trees or lots of fewer larger trees. When timber production is the objective, larger individual trees are generally preferred since this gives higher proportion of usable timber, and reduces transport, logging and processing costs per unit timber.

The climatic factors like rainfall pattern, temperature, edaphic factors like soil fertility status, soil type and structure, water holding capacity, etc., latitude aspect of the site that affects availability of incoming solar radiation are also important factors to decide the tree population.

Solar radiation is the most critical limiting factor for production of agricultural crops in an agroforestry system. Generally shade tolerant annual crops like turmeric, ginger, pineapple, etc. allow higher tree density than the less tolerant crops like cereals and pulses.

Management and maintenance operations like movement of equipments and machines in the field may be affected if the tree density is high. The routine operations like fertilization, hoeing, watering, sowing and harvesting of agricultural crops may also be affected. A careful planting arrangement can avoid these problems to some extent.

The personal preference is one of the most important considerations for selecting the optimal tree density in an agroforestry system. A grower may prefer to have a 'parkland' environment by having a small number of trees per unit area whereas another may prefer a 'forest' appearance for his land with higher tree density. The advantage of agroforestry is that there is very little 'waste' of land even with very low tree density because less number of trees per unit area generally means higher agricultural production.

Thinning

Thinning is a useful silvicultural tool to manipulate the tree growth and form once the agroforestry system is established. It simply involves the removal of selected trees from the site. Thinning allows the grower to maintain the tree population within the limits acceptable from the point of view of the associated crops. It is also an opportunity to harvest some products from the trees, decide on the arrangement of the trees, control the shade they cast on the crops and facilitate animal management. Thinning allows the use of a greater share of the limited resources of the site like solar energy, soil moisture, plant nutrients and space for the remaining trees and associated crops. In addition to the density and arrangement of the trees, thinning takes into account the form and health of the trees, eliminating the unhealthy, crooked or bifurcated trees. If malformed or undesired trees are eliminated before they start to compete with the healthy and straight trees, the growth of retained trees will be unaffected and total production will increase. Early thinning of trees is also required to reduce the competition between the trees and agricultural crops in the agroforestry system, thus increasing the production of arable crops. The specific objectives of thinning of tree crops may be summarized as follows.

1. To have the desired tree density in the system
2. To improve the hygiene of the crop by removing dead, dying, insect infested and diseased trees
3. To ensure the best physical conditions for growth of both tree crops and agricultural crops
4. To obtain the desired type of tree crop from the system
5. To save marketable volume of wood that will be used as fuelwood, pole, staking material, etc. from anticipated loss
6. To improve the quality of wood
7. To increase the net yield and financial return from the system per unit of land
8. To meet the immediate requirements of the grower

Thinning should be done at a time when the resultant debris will cause the least damage to the underneath agricultural crops. Thinning operation should ideally be taken up immediately after harvesting or before planting of an annual crop or after heavy grazing to reduce the loss on agricultural production. Thinning of tree species whose foliage has some degree of fodder value, should coincide with periods of high fodder demands. The degree of thinning depends on the number of trees initially planted, the desired number of trees in the system and the major objectives of the agroforestry system. The resource requirements, characteristics and value of the associated agricultural crops also decide the degree of thinning of trees.

Pruning

Pruning is the removal of plant parts to improve the form and growth of trees. It is probably the most important silvicultural tool in agroforestry. Branches are removed with minimum damage to cambium or growing tissue so that the wound will close in the shortest period of time and with the least possibility of wound infection. The major objectives of pruning of timber trees in a block plantation are for training the plant, maintaining plant health, improving the quality of flowers, fruit, foliage and stems, and controlling growth. However, the principal objective of pruning in agroforestry systems is to reduce the shade on the underneath crops. Timber trees in agroforestry systems tend to branch more than those in block plantations due to the wider spacing and reduced lateral competition, and therefore require more frequent and intensive pruning. However, not more than 30% of the crown should be pruned in any case. The season for pruning depends on the seasons for tree growth and development and on the climatic conditions of the site. It is better not to prune during the dry season when the crops need protection from the summer sun.

In general, deciduous trees and shrubs should be pruned when they are dormant, preferably in early spring just before the start of growth. At this time, wound healing will begin almost at once and it will be the most rapid. Dormant pruning will have less detrimental effect on the growth of trees than pruning during active growth of trees. Another advantage of dormant pruning with deciduous trees is that it is easier to select branches which should be removed when the leaves fall. Pruning of dead branches can be done at any time of the year since no living tissue is affected. When removing diseased material, pruning tools should be disinfected between each cut.

Pruning of timber trees, especially in agroforestry systems, can make a significant difference in the quality of the timber and will provide additional products such as firewood, posts and fodder for the farmer. Pruning is really the best preventive maintenance a young plant can receive. Formative pruning is designed to produce a single, straight stem. It should be carried out when the branches and forks are light enough to be removed with secateurs or a pruning knife. Formative pruning is generally carried out while the trees are still young.

Trees grown at wide spacing tend to develop heavy side branches. This varies greatly with species and provenance. Side branches form knots in the timber, the bigger the knots and the more their number, the lower will be the value of the timber. Pruning of the side branches should be done without damaging the bark of the trunk. Careful pruning to remove side branches will, thereafter, produce knot-free timber at that level as the tree stem expands over time. Branch bases left projecting from the trunk produce dead knots in the wood, which reduce timber quality and may also provide entry points for disease and infection causing pathogens, insects or termites. Pruning of lateral branches has no effect on the straightness of the trunk of a tree. However, removal of too many branches at one time may retard the growth of the tree.

Ideally, branches should be removed when they are of a size which allows them to be cut cleanly with a pruning knife or a pair of secateurs. The need for pruning, the intensity and frequency, varies from species to species and also depends on the nature of the associated crops.

The most common types of tree pruning are crown thinning, crown raising and crown reduction. Selective removal of branches on young trees throughout the crown is called crown thinning. This promotes better form and health by increasing light penetration and air movement. Strong emphasis is given on removal of weak branches. This is generally not practised on mature trees. Crown raising is removal of lower branches on developing or mature trees. On the other hand, removal of larger branches at the top of the tree to reduce its height is called crown reduction. It is different from topping because branches are removed immediately above lateral branches, leaving no stubs. Crown reduction is the least desirable pruning practice. It should be done only when absolutely necessary. The new techniques of tree pruning are fixed-lift pruning, variable-lift pruning, selective pruning and tip pruning.

Fixed-lift pruning: It is the complete pruning of all branches below a prescribed point on the stem. The lift is specified as height from the ground and the choice of height is based on the shape of development attained by the stand.

Variable-lift pruning: It is the complete pruning of all branches below a prescribed variable point on the stem. This point may be specified either as a proportion of the height or as a diameter limit. The specification is often directed towards achieving the removal of a consistent fraction of green matter from the trees, regardless of size.

Selective pruning: It is the removal of some of the branches on the stem at various levels above the ground according to a particular prescription. It is used to correct incipient faults in the crown or to remove branches which may be difficult in subsequent pruning. It promotes a deep crown of fine branches and good apical dominance.

Tip pruning: It is the pruning of a branch at a point other than its junction with the stem. Tip pruning is used to retard the development of a branch.

Coppicing and pollarding

Coppicing and pollarding are techniques which can be used in managing certain types of trees. Coppicing involves cutting the tree species close to ground level to produce new shoots from the stump. It normally sends up a number of shoots instead of the original single stem. Coppicing is a useful practice when it is desirable to produce large quantities of foliage close to ground level, as is the case of many fodder species and for leaf extract products. In some species coppicing occurs naturally if the trees are damaged. Pollarding involves cutting back the crown of a tree in order to harvest wood and to produce regrowth beyond the reach of animals and/or to reduce the shade cast by the crown. It is commonly practised during the periods of extreme forage shortage to produce emergency fodder without killing the tree or greatly reducing its height. Vertical pollarding is the close pruning of branches along the stem. Regrowth after coppicing and pollarding is vigorous because the tree's root system has already been well established. *Acacia catechu, Albizia lebbek, Anogeissus latifolia,* etc. are considered as strong coppicers. *Aesculus indica, Chloroxylon swietinia, Hardwickia binata,* etc. are good coppicers and *Adina cordifolia, Bombax ceiba,* etc. are bad coppicers. *Grewia optiva, Hardwickia binata, Morus alba, Salix* spp., etc. are suitable for pollarding.

Lopping

Removal of one year shoots or fresh growth from entire crown of the tree/plant in order to get sufficient fodder for livestock is known as lopping. Lopping is extensively done in *Morus, Grewia, Bauhinia,* etc.

Bending

Restricting the development of bole to allow more food material to new leaf shoots is called bending. Bending is useful when it is desirable to produce large quantity of foliage close to ground level.

Bushing

Bushing is the horticulture operation commonly used to increase fruit production at a convenient height for harvesting.

AGRICULTURAL CROP MANAGEMENT

Agricultural production in an agroforestry system may come from various crops like annual agricultural food crops, agricultural cash crops, horticultural crops, pasture for grazing, forage crops for harvesting of fodder, etc. or any combination of these. The selection of compatible crops is as important as the choice of tree species. However, certain principles should be followed while choosing agricultural crops to be grown with commercial trees in agroforestry systems.

1. The crops should not produce dense shade before the trees are well established. For example, cocoa should not be planted before a timber species.
2. Climbing species such as yam, vanilla and pepper should not be planted while the trees are young.
3. Crops that compete severely for nutrients and water should not be planted during the tree establishment stage.
4. Crops such as banana, maize and sugarcane that exhaust soil nutrients can negatively affect the establishment of trees on the site if adequate fertilizers are not applied.
5. Root and tuber crops with extensive horizontal root systems should be planted at a sufficient distance from the tree roots to avoid competition and damage during harvest of root crops.
6. It is an advantage if the crop fixes nitrogen. Annual legumes should ideally be included in the crop rotation.

7. The crop should not have the capacity to behave like a weed.
8. The crop should not act as a host to pests and diseases that can affect the trees.
9. In permanent agroforestry systems the crop must be shade tolerant (like ginger, turmeric, arrowroot, pineapple, etc.) or shade demanding (like cocoa) when the tree canopy closes.
10. The crop should require minimum manual labour when it is intended to promote reforestation programmes. The manual labour required for the agricultural crop can limit the total area that can be reforested in agroforestry systems.
11. The crop should have good economic potential.
12. The crop should not cause physical damage to the trees. For example, felling bananas in a new tree plantation can cause damage.
13. Perennial grasses should not be planted before the tree establishment.
14. The crop should not have allelopathic effects on the trees.

Due to the presence of perennial trees, the nature of system continues to change throughout the rotation of the system. Thus, agricultural management practice is also varied during the development of the tree crops both to minimize damage to the trees and to maximize agricultural production. On the basis of tree growth and management, the development of an agroforestry system is divided into five stages such as preplanting phase, establishment phase, silvicultural management phase, maturing phase and harvesting phase.

Preplanting phase

This is the phase where farmer prepares a site for the establishment of agroforestry system. This involves selection of site, crop species and varieties, collection of good quality seed, careful land preparation, and proper management of rodents, insects, white ants, etc. Farmers are encouraged to use locally adapted varieties to ensure good crop establishment and high yield. In this phase the site is prepared and fertility of the site is increased by growing one or two annual legume crops before planting the trees. This will improve soil structure, controls weeds and prepare the land for plantation of trees.

Establishment phase

In this phase of management, the young trees are carefully tended until they are large enough to fend for themselves. Planting arrangement of trees, rooting patterns of trees and agricultural crops, duration of the crops and growth habits of both trees and crops decide how close crops can be grown to the trees. Arable crops should not be planted in close proximity to the base of trees in this phase because this increases competition between the trees and crops for nutrients and water, and will adversely affect the tree establishment and growth. To encourage the early establishment and root development of the trees, a metre-wide free zone should be left on either side of the tree row. The agricultural operations which retard the tree growth even for a short period should be avoided as far as possible.

Silvicultural management phase

Thinning and pruning are the two important silvicultural operations carried out in this phase of management. By the age of first thinning and pruning operation, the trees begin to compete with the agricultural crops on the site. Light becomes the most limiting factor to affect the understorey crops as the canopy closes. To allow light for these crops, the side branches of trees should be pruned. While thinning or pruning the trees, care must be taken not to cause any type of damage to the agricultural crops. Once the productivity is reduced due to shade effect, shade tolerant crops like ginger, turmeric, arrowroot or pineapple, etc. should be grown. Pasture crops may also be grown.

Maturing phase

Different management practices are implemented beneath the maturing trees in this phase of management, such as grazing, maintenance of moisture and nutrients and weed control, which improve and augment the aesthetic value of the stand.

Harvesting phase

In this phase of management tree crops are harvested. Timber harvesting from an agroforest should begin as soon as trees become large enough to fetch a suitable or profitable price. The optimum time for harvest depends on the value of the standing timber and its expected increase in value over time, availability of a suitable market then and in the near future, the owner's preferences. The shape, size and diameter of the timber and the cost of felling of tree and transporting it to the point of sale determine the value of the timber.

MANAGEMENT OF FRUIT TREES IN AGROFORESTRY SYSTEMS

Perennial fruit crops can be used for almost any objective that trees might be used for in agroforestry systems, depending on the needs of the farmer, the nature of the fruit species, and the other components of the system. Fruit trees may be used as hedges, individual trees or live fences. They may also be found in homegardens or as service trees among perennial crops. The fruits are important for domestic consumption and are also sold when a market can be found. However, certain criteria must be followed for the selection of fruit trees for agroforestry systems.

1. The crop should have a commercial potential at the local and regional level.
2. Its growth habit must be compatible with other crops or trees in the system, for example a crown that allows light to reach the associated crops.
3. Fruits should have the ability for prolonged storage and post-harvest processing and transport.
4. The individual fruit tree should have the potential for high production of fruits or biomass.
5. Fruit trees should not act as a host to pests and diseases that are a threat to the associated crops.
6. These should have deep rather than surface root system.
7. Fruits should have the desirable characteristics for industrial uses, for example pectin content for jelly making.
8. Fruit crops should possess the easier vegetative propagation methods.
9. Fruits should have good flavour and high quality.
10. There should have easy fruit harvesting methods.
11. Fruit crops should have short period of establishment.
12. There should be low incidence of pests and diseases.
13. Fruit trees should have multiple uses.

Optimizing the use of light, water and nutrients during the formation and development of the fruit increases the quantity and quality of the fruit produced. Therefore, treatments to improve the capture of solar energy and application of water and plant nutrients will maximize the number of fruits that reach maturity. Abortion of fruits and flowers occurs when the inflorescence does not receive sufficient resources.

It is important to prune and train fruit trees in agroforestry systems. This reduces competition for nutrients, water and light with the associated crops and competition among branches for the tree's resources. Pruning also promotes regular flowering and fruiting, and maintains a low crown that facilitates fruit harvest and the control of disease. Old, sick, damaged and diseased branches and sprouts should be pruned regularly.

CONSTRAINTS IN ADOPTION OF AGROFORESTRY TECHNOLOGY

Institutional constraints

All the forest lands including hilly, deforested and degraded lands that deserve rehabilitation through agroforestry systems are under the jurisdiction of state forestry departments. The forestry officials stick to the classical forestry concept and regard agroforestry systems as incompatible. They believe that farmers' participation is neither suited nor needed and they reinforce this view by depriving farmers of forest land tenure and equitable share in forest benefits. These policies prevail in even where government does not have the potential to rehabilitate such lands by itself. People's participation through agroforestry practices could be a potent means of restoring both protective and productive woody vegetation in barren areas.

Government policy related constraints

In India, there is no well defined agroforestry policy either by the state governments or the central government. Even there is no specific policy for felling of trees. Very often the private growers are not allowed to harvest the trees from their own lands at their need which discourage strongly to go for tree cultivation. A clear-cut government policy is also lacking for inter-state transport of forest products including timber, small timber and other minor forest products. Although tree farming requires high initial investment and return is usually delayed there is no policy for financial support to the tree growers or agroforesters through nationalized banks.

Sociocultural constraints

Majority of farmers in developing countries own or cultivate small sized farms. Their immediate priority is food production from each inch of land. They resist displacing food crops with trees. Farmers prefer only high utility perennial species like bamboo and coconuts. Agroforestry systems are also very labour intensive which may cause scarcity at times for other farm activities. Farm families have traditionally developed labour strategies to use family members at various times of the year for different tasks. Thus, they resist changes in the labour practices of the farming system into which they are introduced.

Socioeconomic constraints

Social acceptability of agroforestry is very closely linked to the economic feasibility of the system. Direct and immediate income that can be derived from a land-use system will be an important criterion in the appraisal of its social acceptability. However, a longer period is required for trees in an agroforestry system to grow to maturity and acquire an economic value. The traditional farmers also do not prefer agroforestry as it requires high initial investment and risk factors are involved for economic returns.

Market related constraints

There are no adequate wood based enterprises with low to medium range investments which affect the farmers the most. Marketing is a big issue for forest products as no privilege is allowed in tree marketing like in case of agricultural marketing. Usually minimum support price for the tree products and other forest products is not fixed by any government agency.

Environmental constraints

Trees serve as hosts to insect pest and diseases that are harmful to agricultural crops. Increased susceptibility to pests and diseases often leads to dependence on potentially harmful pesticides. Trees in agroforestry systems often compete with agricultural crops for light, water and nutrients from the soil.

Technological constraints

Agroforestry system is very difficult to manage and needs more accuracy with highly skillful management practices. Agroforestry is more complex, less understood and more difficult to apply as compared to monocropping.

Chapter 8

Agroforestry Diagnosis and Design

In agroforestry systems the various components like tree, crop and pasture exist in different proportions and orientations. It is difficult to find out which agroforestry system is the best suited for a given land situation. Similarly, it is to be decided which technologies are required for refinement and improvement of the existing agroforestry practices. But without sufficient knowledge of the existing system in a particular land situation, it is very difficult to set the research priorities for modification and development of this system.

Diagnosis and Design (D & D) is a systematic and objective methodology developed by International Centre for Research in Agroforestry (ICRAF) to initiate, monitor and evaluate agroforestry programmes. D & D is based on the philosophy that knowledge of the existing situation (diagnosis) is essential to plan and evaluate (design) meaningful and effective programmes in agroforestry research for development. The methodology plays a strategic role in all the phases of the agroforestry research process. D & D in agroforestry is unique and it has been specially developed for the following purposes.

1. To describe and analyze existing land use systems and their constraints
2. To design appropriate agroforestry technologies to address those constraints
3. To design appropriate research work such as trials and surveying

The basic unit of D & D analysis is the land use system (LUS). The LUS can be defined and analyzed at the level of a country, ecozone, farming system or any other unit. The structure and function of any LUS are determined by climatic, physical, biological, technological, economic, social and political factors. D & D focuses on the interactive effects of these factors on the LUS and searches for opportunities for improvement and development in the LUS.

KEY FEATURES OF D & D

Flexibility: D & D is flexible in its procedure which can be adopted to fit the need and resources of different users.

Speed: D & D has been designed with the option of a 'rapid appraisal application' at the planning stage of a project with in-depth follow-up during project implementation.

Repetition: D & D is an open-ended learning process. There is always a scope for improvement of the initial designs. Thus, the D & D is not completed until further improvements are no longer required.

Diagnosis and Design is done at two levels- macro and micro level. Macro D & D is a large scale analysis of an ecozone within a country. It is important for deciding agroforestry research agenda and extension policy of a country at the national level. But micro D & D focuses on the land use system within a larger ecozone that has special priority for agroforestry intervention. Micro D & D involves a detailed analysis of production systems of households in the LUS which in turn guides for research that will address the constraints of the prioritized LUS.

AGROFORESTRY SYSTEMS RESEARCH PROCESS

The basic objective of agroforestry research is to develop technologies to solve farmers' problems in land use systems in specific ecozones. An agroforestry technology should be defined with reference to at least its main components like multipurpose tree species, spatial arrangement, management and performance levels. ICRAF has developed a research process that uses a systems perspective and an interdisciplinary approach. The process is called 'Agroforestry Systems Research'. There are six main steps in the process.

1. Macro D & D (national and ecozone level)
2. Micro D & D (land use analysis at the production systems level)
3. Technology design
4. Component experimentation
5. Technology testing
6. Technology dissemination and adoption

Macro D & D

There are four main objectives of macro D & D.

1. To identify broad issues and problems constraining all the land use systems in a given ecozone
2. To identify and prioritize areas for potential agroforestry interventions
3. To identify research priorities and formulate research programs
4. To identify needs, opportunities and mechanisms for inter-institutional collaboration for technology development

To meet these objectives, macro D & D uses rapid appraisal techniques. It relies heavily on secondary data which are verified and complemented by quick field surveys. There are seven steps in macro D & D exercise.

1. Identification of study ecozone
2. Delineation of land use systems within the ecozone
3. Description of land use systems
4. Analysis of land use system constraints and potentials
5. Analysis of potential agroforestry technologies
6. Definition of agroforestry research needs
7. Inter-institutional coordination

Identification of study ecozones
The first step in macro D & D is the selection of an ecozone for study. The zone covered in a macro D & D exercise is large, containing significant variations in land characteristics with respect to current uses and constraints. The choice of the study zone should reflect its biological and socioeconomic importance at the national level based on the following factors.

i. Zone's contribution to food production and/or income
ii. Total population it supports and/or the area it covers
iii. Urgency of its constraints
iv. Extent of its unexploited potential for production
v. Level of its development with respect to other areas

Delineation of land use systems
A land use system within an ecozone level is defined as a "population subgroup in which the features and constraints of the farming systems are sufficiently homogeneous to yield similar results if a given agroforestry technology is introduced into those farming systems". The main guideline for distinguishing land use systems is that each system should display unique constraints and potentials differentiating it from other systems in the ecozone of interest. Thus, an LUS consists of a distinctive combination of trees, crops, livestock, soils and other production systems occupying a given unit of land where specific outputs are desired and obtained by a given management unit. Normally the smallest unit of decision-making is the household, but any unit that makes management decisions collectively and shares intimately in the input-output flows of a system is also considered to be an LUS unit.

Description of land use systems
All delineated LUSs are described by specifying the characteristics that are known to affect their current management and performance and would be

expected to affect the introduction of potential agroforestry technologies. These characteristics include title of the system and its location with administrative and political divisions; ecological characteristics like topography, rainfall pattern, soil factors and vegetation; socioeconomic and land use characteristics; resources and supporting services available; and development activities and policies adopted.

Analysis of land use system constraints and potentials
Each system has to be evaluated for factors that prevent its households from obtaining optimal outputs from the available resources. This step requires analysis of farmers' needs and priorities to see how well these are being met by current performance of the LUSs. The performance gap is evaluated by comparing the present levels of outputs with the biophysical and socioeconomic potential of the resources. The range of yields obtained in different LUSs should be compared with yields obtained in on-station or on-farm research. Constraints analysis is based on problems facing households and emphasis is put on constraints which agroforestry can address. To diagnose constraints properly, the relationships between manifested symptoms and causal factors must be understood clearly. The team must be able to interpret the relationships between these factors and the objectives of the household. Furthermore, it must be able to determine what opportunities exist to address the constraints. Constraint analysis is done concurrently with LUS characterization. For example, if one observes steep slopes in cropland, one can conclude that soil erosion is a likely hazard if nothing is being done to prevent it.

Analysis of potential agroforestry technologies
Potential interventions are identified and assessed for their relevance to the constraints and their likelihood of increasing or sustaining productivity of the LUSs. First, all possible interventions are identified, whether they are from the areas of agronomy, forestry or agroforestry. For example, low soil fertility could be addressed by various technologies such as fertilizer, livestock manure, green manure from trees or shrubs, crop rotations. Next, each alternative is evaluated for its technical potential and suitability to farmers' resources and capabilities, infrastructure and support services. A judgement is then made on what interventions seem to have the highest potential. Agroforestry interventions are proposed only when they have a comparative advantage.

'*Ex-ante*' evaluation of a technology is part of technology assessment. It is carried out to determine a technology's potential for adoption. *Ex-ante* evaluation means the evaluation of the likely impact of a proposed technology before the technology has been introduced. It is based on appropriate assumptions using relevant data from other sources. This requires knowledge of technology management and performance under the specific conditions of the LUS.

Definition of agroforestry research needs
If a proposed technology is well known and some farmers are familiar with its management and requirements, then a recommendation for extension programme can be formulated. On the other hand, if very little is known about the technology, then the D & D team is to propose research activities. This research will address critical information gaps for designing viable and adoptable technologies. The team should next carry out a comparative analysis of the research needs for each agroforestry technology for each LUS within an ecozone. This analysis will be the basis for the design of appropriate research programme. Thus the main output of the macro D & D exercise is the definition of a research agenda to develop relevant technologies for the ecozone of interest.

Inter-institutional coordination
A macro D & D exercise should initiate an inventory and review of past and present agroforestry research or development programmes. It will also suggest specific problem areas for complementary research in different institutions and better use of their scientific and physical resources. Multi-institutional participation in strategic phases of the research definitely facilitates the integration of individual efforts and the development of coordinated programme.

MICRO D & D

The major difference between macro D & D and micro D & D is that the former has a broad scope (*i.e.*, an ecozone) whereas the latter focuses on detailed analysis of one prioritized LUS. There are three main objectives of micro D & D.

1. To describe and analyze an LUS in order to identify its constraints
2. To design and evaluate agroforestry technologies to address the constraints
3. To design and evaluate appropriate research programme aiming to develop these technologies

The basic principles for achieving these objectives are similar to that of macro D & D. The choice of the LUS for micro D & D depends mainly on three criteria.

1. Political and economic importance of the system
2. Technical potentials for improvement of the LUS
3. Scientific expertise and other resources in the national collaborating institutions for carrying out research in the LUS

Analysis of land use system and constraints

This phase of micro D & D aims at prioritizing the needs of the household, identifying production constraints and assessing potentials for system development. The basic framework used for this analysis is a farming system, where the decision-making unit is the household. The household usually manages a combination of crop, livestock and tree production systems, along with other non-agricultural and off-farm activities to satisfy its basic felt needs of food, fodder, cash, fuelwood, timber and security. Besides endogenous factors, the farming system is influenced by exogenous factors of a political, social, economic or technological nature. Understanding the interactions within the farming system and the effects of environment is essential for defining the aims of the micro D & D. Thus, the research team has to quantify the resources, management and yield of each component of the farming systems, including characteristics and priorities of the household. The constraints analysis aspect of micro D & D is well suited for planning the initial stages of research. However, later stages of research may require a reassessment or a more precise measurement of some of these constraints.

Design and evaluation of agroforestry technologies

The design and evaluation objective of micro D & D focuses mainly on technology specification and *ex-ante* evaluation of technology. Technology specification for any type of production system refers to a 'package' of husbandry practices and inputs which is specified in terms of the farming systems or households; its components and resource requirements; the management and implementation regimes to be followed by the farmers; and the estimation of real benefits and costs to the farmers under favourable and unfavourable conditions. In other words, technology specification should provide sufficient detail to permit technical feasibility analysis, socioeconomic analysis and assessment by farmers.

Technology specification demands a lot from the D & D team. It raises a large number of specific questions requiring knowledge of the farming systems and scientific expertise. If the questions raised cannot be resolved to the satisfaction of the team, then specific priorities for research have to be established. Because agroforestry is a relatively new science, there is a dearth of technical information on most components. For example, information on the biophysical productivity of multipurpose trees (MPTs) under different arrangements and management regimes is known only for a few species in selected environments. Successful technology design requires adequate research experience. The D & D team may design several technologies to address constraints of the farming system, in terms of the management requirements and performance levels.

Once a technology has been specified in its main components, it becomes possible to carry out an *ex-ante* evaluation based on data from relevant situations. *Ex-ante* evaluation of a technology is simply the analysis of its probable impacts and implications. This analysis looks at benefits and conflicts or problems likely to arise at the levels of farming system, community or village and region or catchment area. It should assess the production potential and technical feasibility of the technology. Economic viability, sustainability, farmers' acceptability and adoption potential of the technology are the four essential types of analysis required in *ex-ante* evaluation. The larger and more complex the technology, the more demanding is the *ex-ante* analysis.

Design and evaluation of research programme
There are four types of scientific research.

1. Basic research which is designed to generate new knowledge or understanding
2. Strategic research to solve specific research problems
3. Applied research to create new technology
4. Adaptive research to adjust technology to the specific needs of a particular set of biophysical or socioeconomic conditions

These are part of a continuum in the technology development process. Productive research requires an integrated and complementary research strategy consisting of on-station research and on-farm research. On-station research consists mainly of basic and applied research and it must be able to offer technical components, information and support to the on-farm research activities. On-farm research complements but does not substitute for on-station research. On-farm research provides feedback for setting on-station research priorities and adapting technologies or components coming out of on-station research. In the case of agroforestry, where basic and applied research is not well developed and where farmers have more experience than scientists with management of technologies, on-farm research may have a stronger role to play in the research strategy including applied research.

Research design criteria
A technology comprises a number of components. Experiments are designed to develop the technical components and to understand relationships among them. Agroforestry technology must be specified at least in its principal components like MPT species, spatial arrangement, management regimes and performance levels. Different types of trials are conducted to achieve these specifications. Within the D & D scheme of ICRAF, the three general categories of agroforestry trials are multipurpose tree species screening trials,

MPTs technology screening trials and MPTs management trials. While all D & D teams have a mandate to design a programme of research, the scope of their proposals may vary. To achieve this, it may be necessary for the D & D teams to carry out multi-visit surveys and interactions with land users and to review scientific secondary information.

METHODOLOGICAL CONSIDERATIONS IN D & D

Research team

The nature of agroforestry systems suggests the need for setting up of multi-disciplinary and multi-institutional teams to achieve the objectives of D & D. For macro D & D, the research team comprises 5 to 10 scientists from biophysical and socioeconomic fields including agronomy, soil science, horticulture, plant protection, climatology, animal science, forestry, agricultural economics, rural sociology or anthropology. These scientists should have experience in research and extension. For micro D & D and follow-up studies, the expertise and composition of the team are based on the prioritized LUS, its identified constraints and the specific objectives of the study. Team leadership is critical for successful D & D implementation. Interdisciplinary interactions and specific contributions made by each discipline are used to formulate the overall research strategy. An objective D & D strategy is the key to bring various disciplines to work together to address the problems of the farmer in an integrated manner. Care is taken to provide an effective interface between the D & D exercise and the planning and implementation of technology development research.

Research domains and recommendation domains

Defining the target land use system or farming system is probably the most crucial step in the 'agroforestry systems research' process. Unless researchers have developed the technology and know how to manage it, what requirements it has and exactly what it can do, they do not have a solid basis to define an appropriate recommendation domain. Thus, in the preliminary stages of technology design, it is better to speak of a 'research domain' until there is sufficient understanding of the technology to determine precisely where it can fit. Accordingly, the research team should always be concerned about whether and how technology development is modifying the original research domain.

Precise definition of the target system is also important for the reason that every farming system or household is somewhat different. Some customizing or fine-tuning of technology is required for each case during the adoption process. Thus the definition of research domains should strike a wise balance not too general so as to be useless as a guide to research, nor too specific so as to apply only to a small number of farms or households.

The research team can assess the effect of its work on the research domain by answering the following questions.

1. Which LUSs are most affected by the problem under investigation?
2. Which systems can benefit and to what extent can they benefit from the technologies being developed?
3. What are the major conditioning and determinant factors, endogenous or exogenous to the farming system, for technology management and performance?
4. What are the expected benefits and impacts of technology adoption?

The concept of research domain is a tool to facilitate and expedite the task of focusing on these key questions. If the team can answer them adequately, the concept has served its purpose.

Data collection methods

The D & D methodology employs several data collection methods appropriate to its specific objectives. Each method has its own strengths, weaknesses, degree of reliability of collected data and resource requirements. For example, the informal survey is fairly effective for identifying constraints, designing technology, promoting interdisciplinary interactions and contributing to research planning, but the reliability of the data is not up to the standard of other methods. D & D exercises should generate a minimum of raw data, a maximum of useful information and in a timely matter. For this reason, the preliminary D & D work will take a rapid appraisal approach using secondary information surveys. D & D work at later stages in the research and development process will use methods that make a maximum contribution within the limits of available human and physical resources.

Analytical methods and the role of farmers

The major decisions in the D & D analysis are derived using interactive and holistic methods with the principal actors being the D & D team and the farmers or households. This interaction should be based on solid information, consultation with development agents and policy makers, and a commitment to arrive at logical conclusions in the process. To ensure effective participation in discussions, all participants must show mutual respect and accept that each can make a valuable contribution. However, this does not mean that participants should accept everything said. There is a need to challenge, seek clarification and discover the root causes of disagreements or conflicts. In this respect, the farmers should be treated as equal participants.

Logistical and operational aspect

A macro D & D is usually completed in about three months. This includes two to three days to plan the study and orientation of the team, two to three weeks for review and synthesis of secondary information, two to three weeks to conduct field work and four to six weeks to analyse the information and prepare the report. The review work and report preparation do not require participation of the whole team; only two members of the team can do these with occasional assistance from the others.

For micro D & D, the total time requirement and logistic support are approximately similar. However, a formal survey of 5-100 farmers with a semi-structured questionnaire constitutes an important part of the study. In addition, the team may need to allocate relatively more time to the review of relevant agroforestry researcher and extension work to strengthen its analysis.

Chapter 9

Agroforestry Policy and Projects

Trees have been used in cropping systems since the beginning of agriculture. Throughout the world, at one period or another in its history, it has been the practice to cultivate tree species and agricultural crops in intimate combination. In the tropics, human beings underwent a transition from hunting/gathering to the use of domesticated plants and livestock. As a part of the process they cut down trees, cleared the debris by burning and sowed crops in the ash-enriched soil. It was the 'slash-and-burn' agriculture, a primary forerunner of the present day agroforestry and a practice that might have originated in the Neolithic period, around 7000 BC.

There are innumerable examples of traditional land-use practices involving combined production of trees and agricultural species on the same piece of land in many parts of the world. Trees were an integral part of these farming systems and they were deliberately retained on farmlands to support agriculture. These practices are now known as agroforestry. However, the ultimate objective of these practices was not tree production but food production.

HISTORY OF AGROFORESTRY RESEARCH

By the end of the nineteenth century, establishing forest or agricultural plantations had become an important objective for practising agroforestry. But the research on tree plantation was carried out by foresters. The foresters conducting the research never envisioned the system as being capable of making a significant contribution to agricultural development, or its potential as a land-management system.

In 1914, Russell Smith, an economic geographer at Columbia University in New York advocated the use of permanent tree-protected areas to maximize production on arable land. But the post-World War II economic and industrial boom lessened interest in agroforestry. However, a decade later new technologies and innovative approaches to agriculture and land-use systems coupled with increased interest in conservation and concern for degrading natural resources, once again brought attention to agroforestry.

In the 1960s and early 1970s there was increasing concern for the forested lands of the tropics. It was clearly recognized that they were under severe pressure. Some thought that commercial exploitation was the problem; others believed that fuelwood needs were the culprit, while still others thought that shifting cultivation was the root cause. In 1975, the International Development Research Centre (IDRC) located in Ottawa, Canada engaged Mr. John Bene, a retired forest industrialist to study the problem. Bene formed a small advisory committee in Canada, and recruited experts in the various continents to prepare studies pertinent to their area with the following major objectives.

- Identify significant gaps in world forestry research and training
- Assess the interdependence of forestry and agriculture in low income tropical countries and propose research leading to optimization of land use
- Formulate forestry research programmes which promise to yield results of considerable economic and social impact on developing countries
- Recommend institutional arrangements to carry out such research effectively and expeditiously
- Prepare a plan of action to obtain international donor support

Based on these studies the IDRC published a report titled 'Trees, Food and People: Land Management in the Tropics' in 1977. This IDRC Project Report for the first time recommended the creation of an internationally financed Council for research on agroforestry, to administer a comprehensive programme leading to better land use in the tropics. The report suggested that the objectives of such a Council should be

1. The encouragement and support of research in agroforestry.
2. The acquisition and dissemination of information on agroforestry systems
3. The promotion of better land use in the developing countries of the tropics. First priority should be given to the combined production system which would integrate forestry, agriculture, and/or animal in order to optimize the land use.

INTERNATIONAL INSTITUTES WORKING IN FOREST CONSERVATION AND RESEARCH

World Agroforestry Centre (WAC)

With the recommendations of IDRC, an internationally financed organization, then known as the International Council for Research in Agroforestry (ICRAF) was established in 1977 and started its work in The Hague. The organization moved to its present headquarters in Nairobi, Kenya in 1978. In 1992, it was renamed as International Centre for Research in Agroforestry (ICRAF) when it joined the Consultative Group on International Agricultural Research (CGIAR). In 2002 the Centre's name was once again changed to World Agroforestry Centre (WAC) in order to reflect its global reach, as well as its more balanced research and development agenda. However, the Centre's legal name - International Centre for Research in Agroforestry remains unchanged.

The World Agroforestry Centre has since been involved in the promotion of agroforestry research programmes with a heavy emphasis on Africa. Agroforestry research has accelerated rapidly since the early 1980s and has resulted in a greater understanding of the science. Much of this understanding has come from observations of existing practices and systems, although an increasingly important knowledge base is being established through designed agroforestry experiments.

WAC is the only institution that does globally significant agroforestry research in and for all of the developing tropics. The Centre conducts research in agroforestry, in partnership with national agricultural research systems with a view to developing more sustainable and productive land use. The focus of its research is countries/regions in the developing world, particular in the tropics of Central and South America, Southeast Asia, South Asia, West Africa, Eastern Africa and parts of central Africa. Knowledge produced by WAC enables governments, development agencies and farmers to utilize the power of trees to make farming and livelihoods more environmentally, socially and economically sustainable. WAC's work also addresses many of the issues being tackled by the Sustainable Development Goals, specifically those that aim to eradicate hunger, reduce poverty, provide affordable and clean energy, protect life on land, and combat climate change.

Building upon the success of the scientific studies carried out by WAC at its centres, agricultural scientists all over world began investigating the feasibility of intercropping in plantation and other tree crop stands as well as studying the role of trees and shrubs in maintaining soil productivity and controlling soil erosion. Livestock management experts also began to recognize the importance of indigenous tree and shrub browse in mixed farming and pastoral production systems. Many of these studies and efforts provided important knowledge about the advantages of combined production systems involving crops, trees and animals.

Agroforestry is now taught as a part of forestry and agriculture degree courses in many universities in both the developing and industrialized world. Agroforestry research is conducted by universities, government agencies, individual agricultural producers and other organizations. Indeed, agroforestry is fast becoming recognized as a system which is capable of yielding both wood and food and at the same time capable of conserving and rehabilitating ecosystems.

Food and Agriculture Organization (FAO)

The Food and Agriculture Organization (FAO) is a specialized agency of the United Nations that leads international efforts to defeat hunger. Its goal is to achieve food security for all and make sure that people have regular access to enough high-quality food to lead active, healthy lives. FAO works in over 130 countries worldwide with its headquarters in Rome.

FAO plays an important role in coordinating and implementing forestry genetic resources policy within its overall aim of providing technical assistance. Emphasis is paid, in technical terms, on the use of multipurpose species. The species in the arid and semi-arid regions of developing countries are given priority as they are subjected to high human and biotic stress. Efforts are done to explore, use and conserve gene resources of forest trees. It disseminates the information on forest tree seed supplies, seed collection, handling, storage, testing and certification.

The International Board for Plant Genetic Resources (IBPGR)

The International Board for Plant Genetic Resources (IBPGR) is an international scientific-organization under the aegis of the Consultative Group on International Agricultural Research (CGIAR) and established by them in 1974. The basic function of IBPGR is to advance the conservation and use of plant genetic resources for the benefit of present and future generations. It is working on agricultural crops, forest species, particularly fuelwood species, for their conservation and improvement. It has raised general awareness of the issues related to genetic resources, which in turn has stimulated the demand for training and research, for technical publications, and for scientific assistance to national plant genetic resources programmes. The institute has been involved in the collecting of 200,000 samples of crops in 120 countries.

International Tropical Timber Organization (ITTO)

The International Tropical Timber Organization (ITTO) is an intergovernmental organization promoting the sustainable management and conservation of tropical forests and the expansion and diversification of international trade in tropical timber from sustainably managed and legally harvested forests. The major activities of ITTO are

- Develops internationally agreed policy guidelines and norms to encourage sustainable forest management (SFM) and sustainable tropical timber industries and trade.
- Assists tropical member countries to adapt such guidelines and norms to local circumstances and to implement them in the field through projects and other activities.

- Collects, analyzes and disseminates data on the production and trade of tropical timber.
- Promotes sustainable tropical timber supply chains.
- Helps develop capacity in tropical forestry.

ITTO is an action and field-oriented organization with more than 30 years of experience. It has funded and assisted in the implementation of more than 1000 projects and other activities addressing the many aspects of SFM, such as forest restoration, wood-use efficiency, the competitiveness of wood products, market intelligence and transparency in the tropical timber trade and tropical timber supply chains, forest law enforcement and governance, illegal logging, biodiversity conservation, climate-change mitigation and adaptation, the contributions of non-timber forest products and environmental services, and the livelihoods of forest-dependent communities. Its headquarters is in Yokohama, Japan.

International Union for the Conservation of Nature (IUCN)

The International Union for Conservation of Nature (IUCN) is a unique international agency composed of both government and nongovernmental civil society organisations with a mission to influence, encourage and assist societies throughout the world to conserve the integrity and diversity of nature and to ensure that any use of natural resources is equitable and ecologically sustainable. It provides public, private and non-governmental organisations with the knowledge and tools that enable human progress, economic development and nature conservation to take place together. By facilitating these solutions, IUCN provides governments and institutions at all levels with the impetus to achieve universal goals, including on biodiversity, climate change and sustainable development, which IUCN was instrumental in defining.

IUCN is concerned with species level conservation and has paid increasing attention to plant genetic resources. It produces the IUCN plant Red data book, a series that gives detailed case histories on rare and threatened plants in all parts of the world; for each species, data are given on conservation status, threats to survival, distribution and habit, together with a short description and an evaluation of its interest or potential value to humankind.

IUCN's headquarters is in Gland, Switzerland. It has a World Conservation Monitoring Centre (WCMC) located at Cambridge, UK for data storage and processing and also provides information on international trade in endangered plants and animals.

International Union of Forestry Research Organization (IUFRO)

The International Union of Forest Research Organizations (IUFRO) is a non-profit, non-governmental international network of forest scientists. It unites more than 15,000 scientists in almost 700 Member Organizations in over 110 countries with its headquarters in Vienna, Austria. Scientists cooperate in IUFRO on a voluntary basis. IUFRO's vision is of science-based sustainable management of the world's forest resources for economic, environmental and social benefits. It is the global network for forest science cooperation.

IUFRO promotes global cooperation in forest-related research and enhances the understanding of the ecological, economic and social aspects of forests and trees. It disseminates scientific knowledge to stakeholders and decision-makers and contributes to forest policy and on-the-ground forest management. IUFRO has placed strong emphasis on industrial species and works on provenance testing, progeny testing and breeding of specific species. It is also working on conservation and on population genetics for the species of temperate zone and Mediterranean conifers, as well as *Quercus* spp., *Eucalyptus* spp., *Populus* spp.

With the 'Strategy 2015–2019', IUFRO addresses five research themes and associated emphasis areas, and three institutional goals. The following five themes aim to guide the science collaboration within IUFRO's global network in the current Strategy period.

1. Forests for people
2. Forests and climate change
3. Forest and forest-based products for a greener future
4. Biodivesity, ecosystem services and biological invasions
5. Forest, soil and water interactions

The three institutional goals are

- Research excellence: Strive for quality, relevance and synergies
- Network cooperation: Increase communication, visibility and outreach
- Policy impact: Provide analysis, insights and options

United Nations Environment Programme (UNEP)

The United Nations Environment Programme (UNEP), an agency of the United Nations, coordinates the organization's environmental activities and assists developing countries in implementing environmentally sound policies and practices. UNEP's activities cover a wide range of issues regarding the atmosphere, marine and terrestrial ecosystems, environmental governance and green economy. UNEP's mission is to provide leadership and encourage partnership in caring for the environment by inspiring, informing, and enabling nations and peoples to improve their quality of life without compromising that of future generations. It has played a significant role in developing international environmental conventions, promoting environmental science and information and illustrating the way those can be implemented in conjunction with policy, working on the development and implementation of policy with national governments, regional institutions in conjunction with environmental non-governmental organizations. Its headquarters is in Nairobi, Kenya.

Central America and Mexico Coniferous Resources Cooperative (CAMCORE)

An organization called Central America and Mexico Coniferous Resources Cooperative (CAMCORE) was formed in 1980 and made part of the Department of Forestry & Environmental Resources at North Carolina State University, USA. The driving force behind its formation was the private forest sector. AMCORE demonstrates that the private sector, in collaboration with government agencies and universities, can make substantial contributions to tree breeding, gene conservation and environmental stewardship, while simultaneously providing economically sound forest management. This private organization has been able to directly affect the development of genetic resources by the cooperative efforts.

AMCORE makes seed collections in threatened forest stands and then plants these on members' land in more protected areas in genetic field trials (called progeny tests) and conservation areas (called *ex situ* conservation banks or parks) with similar climates in countries around the world. AMCORE professionals at North Carolina State University analyze the data from the progeny trials and produce annual summaries to help members decide on what to grow. Early results from CAMCORE genetic tests indicate that height gains of approximately 20% are possible in the tropics by simply using alternative species and more-productive seed sources. CAMCORE is now working with 50 different forest species and has sampled 11,000 trees in 500 locations. It has more than 2,500 hectares of genetic trials and conservation areas. It has the largest database on tropical and subtropical pines and non-Australian eucalypts in the world.

Tropical Agricultural Research and Higher Education Centre (CATIE)

The Centre Agronomics Tropical de Investigation Ensenanza (Tropical Agricultural Research and Higher Education Centre, CATIE) is an international institute for agricultural development and biological conservation and sustainable use of natural resources in Central America and the Caribbean, combining science, education and innovation. CATIE is the first graduate school in Agricultural Sciences in Latin America. It works for the renewal of natural resources by the renewable natural resources development which has four programmes such as agroforestry, silviculture, wild lands and watershed management. It also works on *ex-situ* conservation.

International Development Research Centre (IDRC)

International Development Research Centre was established by an act of Canada's parliament in 1970 with a mandate to initiate, encourage, support, and conduct research into the problems of the developing regions of the world and into the means for applying and adapting scientific, technical, and other knowledge to the economic and social advancement of those regions. The IDRC funds research in developing countries through various projects which work for forestry research in developing countries to promote growth, reduce poverty, and drive large-scale positive change. The IDRC headquarters is located in Ottawa, Canada. It has four regional offices in Montevideo, Uruguay; Nairobi, Kenya; Amman, Jordan; and New Delhi, India. IDRC is governed by a board of up to 14 governors, whose chairperson reports to Parliament through the Minister of International Development.

Commonwealth Scientific and Industrial Research Organization (CSIRO)

The Commonwealth Scientific and Industrial Research Organisation (CSIRO) is an independent Australian federal government agency with its role to improve the economic and social performance of industry for the benefit of the community. The centre works on Australian forest tree resources. The Tree Seed Centre collects and distributes high quality and source identified seeds of commercially promising Australian woody plants for research purposes. It provides professional advice on the choice of species and seed supply, and also provides technical information on species of value. The CSIRO headquarters is located in Canberra

Danida Forest Seed Centre (DFSC)

The Danida Forest Seed Centre is an institution administered by DANIDA, the Danish International Development Administration, and is situated in Humlebaek, Denmark. DFSC has been in operation since 1969 and provides advice and guidance on seed procurement, tree improvement, and conservation of forest genetic resources for tropical and subtropical developing countries. The DFSC was integrated on January 2004 into the Danish Centre for Forest Landscape and Planning under the Royal Veterinary and Agricultural

University, Denmark. DFSC is currently involved in collecting available information about direct seeding of woody species. Most of the activities of the DFSC are taking place in developing countries, primarily in South-East Asia and Central America. Its major activities include handling, storage and distribution of seeds collected within international projects on genetic resources of arid and semi-arid zones.

Oxford Forestry Institute (OFI)

The Oxford Forestry Institute, formerly known as Commonwealth Forestry Institute comes under the Department of Plant Sciences, University of Oxford, UK. It is a world centre for research and development. Over the last 30 years the OFI has undertaken an extensive programme of research on forest genetic resources of a range of socio-economically important tropical tree species. It is working for the establishment of international provenance testing projects for some 50 species in Central America and parts of Africa. The project covers exploration, taxonomy, collection, seed storage, distribution, conservation and development of genetic improvement strategies for a number of tropical species like *Pinus* spp., *Agattiis* spp., *Cupressus* spp., *Widdringtonia* spp., *Cedrela* spp., *Cordia alliodora*, *Liquidambar styracifua*, *Leucaena* spp., *Prosopis* spp. and *Acacia* spp., etc. These networks of species, provenance and progeny trials have yielded a large body of important data and information on the relative performance of the material as well as patterns of taxonomic and genetic diversity across a range of plantation and agroforestry species.

U.S. Department of Agriculture's Forest Service and State Programmes

The activities of the U.S. Department of Agriculture's Forest Service include research on the biological, physical, and social sciences related to the diverse forests and rangelands. The genetic improvement programmes are also conducted in national forests. Research is primarily aimed at species of high commercial value, but there is growing emphasis on maintaining the diversity of forests. Individual states have independent conservation programmes. Conifer germplasm conservation project is designated to provide information and resources needed for long term protection of the diversity of forests. It is working on *Pinus taeda*, *Pinus ponderosa*, *Pseudotsuga* spp., etc.

AGROFORESTRY RESEARCH INITIATIVES IN INDIA

In India, research work on agroforestry was initiated during the late 1960s and 1970s by the Indian Grass Land and Fodder Research Institute, Jhansi, Central Soil and Water Conservation Research and Training Institute, Dehradun, Central Arid Zone Research Institute, Jodhpur and ICAR Research Complex for the North-eastern Hill Region, Barapani, Meghalaya. However, the organized agroforestry research in India began in May, 1979 when a seminar on agroforestry was organized at Imphal, Manipur by the Indian Council of Agricultural Research (ICAR), New Delhi to accumulate and compile the data on the 'Research and Development' of agroforestry in India.

The All India Coordinated Research Project on Agroforestry (AICRAF) was initiated in 1983 at 20 centres of which 12 were in the State Agricultural Universities (SAU) and 8 at the ICAR institutes. National Research Centre for Agroforestry (NRCAF) was established on 8th May, 1988 at Jhansi, U.P. to undertake basic and applied research for developing and delivering technologies based on sustainable agroforestry practices on farms, marginal and wastelands for different agroclimatic zones in the country. Initially the coordinating unit of the AICRAF was located at the ICAR headquarters, New Delhi. Later on, it was shifted to National Research Centre for Agroforestry, Jhansi in 1997 under the administrative control of the Director, NRCAF, Jhansi with the additional charge of Project Coordinator. The Centre is now upgraded as ICAR-Central Agroforestry Research Institute (ICAR-CAFRI) from 1st December, 2014. At present there are 37 centres of AICRP on Agroforestry located in 27 State Agricultural Universities, 9 in ICAR institutes and one in Indian Council of Forestry Research and Education (ICFRE) institute representing all agroclimates of the country. The agroforestry research through the AICRP on Agroforestry was conceptualized with the following six projects.

1. Diagnostic survey and appraisal of existing farming system and agroforestry practices including farmers' preference.
2. Collection and evaluation of promising tree species, cultivars of fuel, fodder and small timber for agroforestry interactions.
3. Studies on management practices of agroforestry systems.
4. Analyze economical relation of agroforestry systems.
5. Explore the role of agroforestry in environment protection.
6. Studies on post-harvest technology, fishery, apiculture, lac, etc. in relation to agroforestry systems.

The agroforestry research in the country includes diagnostic survey and appraisal of existing agroforestry practices prevalent in different agro-ecological zones of the country. One of the major thrusts of the agroforestry research is collection and evaluation of promising tree species/cultivars of fuel, fodder and small timber. A large germplasm of important agroforestry trees has been collected and evaluated in arboretum established by the AICRP on Agroforestry centres. On the basis of identification and evaluation of promising trees and 'Diagnosis and Design' survey of the existing systems, agroforestry interventions were initiated in different agro-climatic regions.

In addition to ICAR, Indian Council of Forestry Research and Education also conducts agroforestry research through its research institutes and advanced research centres in different parts of the country. Recognizing agroforestry as a viable venture, many business corporations, limited companies such as ITC, WIMCO, West Coast Paper Mills Ltd., Hindustan Paper Mills Ltd., and other institutions initiated agroforestry research with emphasis on production of improved planting material of the fast growing species with an objective to meet the demand of wood based industries (Dhyani, 2018).

NATIONAL AGROFORESTRY POLICY, 2014
Need for agroforestry policy in India

Major policy initiatives, including the National Forest Policy 1988, the National Agriculture Policy 2000, Planning Commission Task Force on Greening India 2001, National Bamboo Mission 2002, National Policy on Farmers, 2007 and Green India Mission 2010, emphasize the role of agroforestry for efficient nutrient cycling, organic matter addition for sustainable agriculture and for improving vegetation cover. However, agroforestry has not gained the desired importance as a resource development tool due to the following factors.

1. Absence of a dedicated and focused national policy and a suitable institutional mechanism
2. Lack of an integrated farming systems approach
3. Restrictive regulatory regime
4. Inadequate attempts at liberalization of restrictive regulations
5. Insufficient research, extension and capacity building
6. Dearth of quality planting material
7. Institutional finance and insurance coverage
8. Weak market access for agroforestry produces
9. Industry operations at a sub-optimal level

Major policy goals

- Setting up a National Agroforestry Mission or an Agroforestry Board to implement the National Policy by bringing coordination, convergence and synergy among various elements of agroforestry scattered in various existing, missions, programmes, schemes and agencies pertaining to agriculture, environment, forestry, and rural development sectors of the Government
- Improving the productivity; employment, income and livelihood opportunities of rural households, especially of the smallholder farmers through agroforestry

- Meeting the ever increasing demand of timber, food, fuel, fodder, fertilizer, fibre, and other agroforestry products; conserving the natural resources and forest; protecting the environment and providing environmental security; and increasing the forest / tree cover, there is a need to increase the availability of these from outside the natural forests

Major policy objectives

The National Agroforestry Policy was launched on February 10, 2014, on the occasion of World Congress on Agroforestry held in New Delhi and India became the first country in the world to have a National Agroforestry Policy. The basic objectives of this policy were

- Encourage and expand tree plantation in complementarity and integrated manner with crops and livestock to improve productivity, employment, income and livelihoods of rural households, especially the small holder farmers
- Protect and stabilize ecosystems, and promote resilient cropping and farming systems to minimize the risk during extreme climatic events
- Meet the raw material requirements of wood based industries and reduce import of wood and wood products to save foreign exchange
- Supplement the availability of agroforestry products (AFPs), such as the fuelwood, fodder, non-timber forest produce and small timber of the rural and tribal populations, thereby reducing the pressure on existing forests
- Complement achieving the target of increasing forest/tree cover to promote ecological stability, especially in the vulnerable regions
- Develop capacity and strengthen research in agroforestry and create a massive people's movement for achieving these objectives and to minimize pressure on existing forests

Strategies for implementation of the National Agroforestry Policy

Establishment of institutional setup at national level to promote agroforestry: An institutional mechanism, such as a Mission or Board is to be established for implementing the agroforestry policy. It will provide the platform for the multi-stakeholders to jointly plan and identify the priorities and strategies, for inter-ministerial coordination, programmatic convergence, financial resources mobilization and leveraging, capacity building facilitation, and technical and management support.

Simple regulatory mechanism: There is a need to create simple mechanisms / procedures to regulate the harvesting and transit of agroforestry produce within the State, as well as in various States forming an ecological region. There is also the need to simplify procedures, with permissions extended on automatic route as well as approval mode through a transparent system within a given time schedule.

Development of a sound database & information system: Security of land tenure is a critical issue. Due to the longer gestation period of trees stability and security of tenure rights is a necessary condition for farmers' to take up agroforestry.

Investing in research, extension and capacity building and related services: The non-existent extension system for agroforestry may be the key reason for non-adoption of technologies. A common web-based platform may be setup to bring all research findings and available knowledge in the area of agroforestry for access of all stakeholders.

Improving famers' access to quality planting material: Certification of nurseries, seeds and other planting materials for agroforestry is required to make available good quality planting material at the required scale. Private sector can play an important role in augmenting supplies of improved planting stock.

Providing institutional credit and insurance cover for agroforestry: There is need for setting up of special purpose banking institutions to address specific needs of agroforestry sector. Agroforestry sector should also be benefitted with the provisions of interest subvention in the line of agricultural credit.

Facilitating increased participation of industries dealing with agroforestry produce: The role of agroforestry/biomass based industries in the promotion of agroforestry is crucial. Therefore, such industry sector needs to be encouraged and facilitated.

Strengthening farmer access to markets for tree products: The marketing system for agroforestry produce is mostly unorganized. Marketing infrastructure similar to what is available for agricultural commodities, including market information be introduced with more private sector participation.

Incentives to farmers for adopting agroforestry: Regional and thematic differentiation in the agroforestry policy needs to be minimized. Incentive and support structure, such as the input subsidy, interest moratorium, etc. during the gestation period for promotion of agroforestry be provisioned to encourage farmers.

Promoting sustainable agroforestry for renewable biomass based energy: Emphasis needs to be on raising fast growing trees/bushes/grasses on marginal and degraded farmlands keeping in view their uses for meeting various energy requirements for making profitable agroforestry practices.

The policy is not only seen as crucial to India's ambitious goal of achieving 33% tree cover but also to mitigate green house gas emissions from agriculture sector. After launching the National Agroforestry Policy in 2014, considerable progress has been made in terms of putting it into practice. Department of Agriculture Cooperation and Farmers Welfare under the Ministry of Agriculture and Farmers Welfare is also playing a significant role in the promotion of agroforestry. It has taken a policy decision to include trees in all its programmes, and this will significantly increase tree-planting on farms, especially under schemes funded by the National Mission on Sustainable Agriculture. Efforts are on to issue guidelines on the production and supply of high-quality planting material and accreditation of nurseries producing planting material (Dhyani, 2018).

Chapter 10

Management Practices of Multipurpose Trees

Multipurpose trees and shrubs are defined as all woody perennials that are purposefully grown to provide more than one significant contribution to the production and/or service functions of a land-use system that they occupy. In agroforestry systems, different species of trees and shrubs can be planted with many types of crops in a variety of patterns. Thus, it is required to know about the methods of propagation and nursery raising, planting and other silvicultural management practices along with the diverse uses of multipurpose trees and shrubs of tropical and subtropical areas before their selection for forestry and agroforestry activities. It is important to select the most suitable tree species since it is not easy to replace them once they have been planted.

MANGIUM (*ACACIA MANGIUM*)

Acacia mangium is a single-stemmed evergreen tree that grows to 25-35 m in height and up to 60 cm in diameter. The bole is usually straight, often fluted near the base, free of branches for up to half its height. Mangium is native to Australia, Indonesia and Papua New Guinea. However, it tolerates varied site conditions and has adaptability to different planting objectives. Mangium shows most vigorous growth on well-drained, fertile soils in high rainfall areas in the humid tropics. It is valued for its rapid growth and has been planted throughout the humid tropics and is a major plantation species in the Asia Pacific. Provenances from Papua New Guinea consistently show better growth in height and diameter.

Propagation and nursery

Mangium starts to flower after 2 years of planting. Pods are broad, linear and irregularly coiled when ripe. Pods can be collected from the trees when the pods turn very dark-green to light-brown in colour. Seeds are extracted manually after sun-drying. Pods and seeds should not be left to dry in the sun for long. The average number of seeds in one kg of pure seed comes about 125,000. The seed can retain its viability for several years when stored at room temperature in a dark place in a sealed container.

Propagation by seedling is best although direct sowing is also possible. To break dormancy of mangium seeds, hot water treatment is recommended. Seeds are treated before sowing by immersing them in boiling water for 30 seconds then soaking them in cold water for 24 hours. Seed inoculation with appropriate rhizobial strain is recommended before sowing. Mangium seedlings are ready for pricking out in 6–10 days after sowing. Polythene bags are the most common containers used in the tropics for pricking out. Mangium seedlings attain a target size of 25–40 cm height in about 12 weeks. Seedlings are hardened by progressively reducing watering and removing shade in the nursery. If the seedlings have grown larger than the target size in the nursery, they may be lopped.

Planting and tree management

Planting is usually done in pits of 20 cm depth and 10–12 cm diameter. In pure stands, spacing of 2 m × 2 m or 2.5 m × 2.5 m is common. In agroforestry situations, spacing within rows and between rows must consider the effect of shade and root competition on the yield of associated crops. *Acacia mangium* demand full light for good development and its growth in shade is stunted. In the 1st year, the plantation should be protected from livestock, as they browse the trees.

First weeding must be carried out two months after planting and thereafter at regular intervals depending on weed growth. Fertilizers may be applied @ 30–40 g N, 15–20 g P_2O_5 and K_2O per seedling per year from the second year to the fifth year. Mangium needs regular pruning and thinning if the plantation objective is to produce quality saw logs on a 15–20 year rotation. These operations in general are not required for pulpwood production on a 6–8

Silvicultural management schedule for mangium saw log regime

Age	Activity	Remarks
Once at 4 months after planting and again 2 months thereafter	General slashing	All climbers within 45 cm radius of each plant are uprooted. All branches at height less than 30 cm from the ground are removed.
12 months after planting	General slashing and first pruning	All branches up to 1.5–2.0 m height are removed.
2 years after planting	First thinning and high pruning	600 trees/ha are maintained and excess plants are removed which may be used as poles. Branches up to 6 m height of the 200 selected trees (to be retained till end) are pruned.
4-5 years after planting	Second thinning	Another 200 trees/ha are harvested retaining 400 trees/ha.

years rotation. In agroforestry systems, frequent pruning is required to allow light for the associated crops. In pruning, branches are carefully removed in one or more steps along the bottom trunk up to about 6–7 m height. For saw log production the following silvicultural schedule is recommended.

Protection

Although root rot disease caused by *Ganoderma* sp., *Phellinus* sp. and *Rigidoporus lignosus* are major problems in mangium stands, there are no specific control recommendations against these fungi. Signs of the disease are evident on the roots after the tree has fallen or upon excavation. The usual method of controlling root rot caused by fungi that spread by root contact is to remove and destroy all diseased roots and woody debris. Chemical protection against pink disease (*Corticium salmonicolor*), especially in endemic areas, can be achieved by using copper fungicides. The best way to prevent pink disease, however, is to plant tolerant varieties.

Progressive decay of the heartwood (heart rot) is another malady afflicting mangium trees. Normally, fungi that decay heartwood do not attack sapwood. Thus, such trees continue to grow to maturity and may outwardly appear healthy and vigorous. However, since heart rot is progressive, there will be considerable decay to the heartwood at the end of the rotation. A variety of basidiomycete fungi have been associated with this malady. At present there are no control measures against mangium heart rot. The best way is to avoid injury to trees and wound dressing.

Although about 30 insect species are reported to be pests of mangium, only a few such as root feeders, branch and stem borers and the red coffee borer are considered economically important. Root feeders (*Sternocera aequisignata*) can be controlled by carbofuran and chlorpyrifos application to the soil or seedbeds. For controlling termites water solution of chlordane 1% or dieldrin 0.5% is applied around the affected area. To prevent branch and twig borer (*Sinoxylon anale*) occurrence, all broken branches in which breeding takes place are removed and burnt. The only effective method to control red coffee borer (*Zeuzera coffeae*) damages is to inject insecticide into the holes where larvae push out their frass.

Uses

Timber is used for a variety of purposes like wood-based panels, pulp and paper industry, etc. Mangium wood gives attractive furniture, cabinets, moulds and door/window components. However, the presence of flutes and incidence of rots and termite attack will detract both the quality and quantity of sawn timber from mangium logs. Therefore, mangium has greater potential as a component of composite wood products such as veneer and plywood, laminated veneer lumber, fibreboards, etc. and for chemical uses such as pulp, paper and tannin production, besides fuelwood. The pulp is readily bleached to high brightness levels and is excellent for papermaking. With a calorific value of 4800 kcal/kg, Acacia mangium provides good quality charcoal and is suitable for the manufacture of charcoal briquettes and artificial carbon.

Mangium bark has high tannin content (18–39%), which can be commercially exploited. Mangium trees form a symbiosis with soil bacteria of the genus *Rhizobium*, leading to root nodules, in which the bacteria transform free nitrogen into organic and inorganic compounds containing nitrogen.

The species is suitable for many agroforestry systems like alley cropping, agrisilvicultural, silvipastoral hortisilvicultural, etc. due to its quick growth and low branching habit. It has potential in some intercropping combinations, such as with maize, pulses, groundnut, turmeric, pineapple, arrowroot, etc.

SISSOO (*DALBERGIA SISSOO*)

Dalbergia sissoo is a medium to large-sized deciduous tree, growing up to 30 m in height and 80 cm dbh (diameter at breast height, i.e., 1.37 m) under favourable conditions. The bole is often crooked and branchless for up to 8 metres, occasionally for as much as 20 m. It develops a long taproot with numerous lateral ramifying roots from an early age. It is native to Afghanistan, Bangladesh, Bhutan, India, Malaysia and Pakistan, and cultivated as a forest tree in south Asia and tropical Africa. Abundant moisture and adequate sunlight are required for desired growth of this plant.

Propagation and nursery

Propagation of *Dalbergia sissoo* takes place most commonly through seed and vegetatively through suckers arising from the root system. There are approximately 45,000–55,000 seeds/kg. Sissoo rarely regenerates under the parent canopy. Ripe pods may be collected manually by climbing trees and picking the fruits or by shaking the branches and picking the pods from the ground. Usually the seeds are not extracted from the pods, but the pods are broken into 1-seeded pieces. Seeds have no dormancy and remain viable for only a few months. Pre-treatment of seeds is not necessary, but soaking in water for 24–48 hours accelerates germination up to 80%. Germination of fresh seed takes 1–3 weeks. Shading is recommended during the hottest hours of the day during the germination period. Propagation by root suckers is done best by cutting stems just below the soil surface. While it is difficult to propagate using stem and branch cuttings without hormone treatments, exogenous application of auxins have been found to improve survival and growth rates.

Planting and tree management

Dalbergia sissoo plantations are established in block or strip plantations at 1.8 m × 1.8 m to 4 m × 4 m. Closer spacing is used for straight timber of good quality. When the canopy closes, at about 6 years, 30–40% of the stems are thinned to selectively remove suppressed, diseased and badly formed trees. Thinning is recommended every 10 years where the rotation is 30–60 years. There is evidence that the stumps begin to lose vigour after 2 or 3 rotations when managed as a coppice crop. It coppices vigorously up to about 20 years of age.

Uses

After teak, *Dalbergia sissoo* is the most important cultivated timber in India. The heartwood is very hard and close grained with a specific gravity of 0.62–0.82. It seasons well and does not warp or split. It is extremely durable and is one of the timbers least susceptible to dry-wood termites in India. Wood offers resistance to sawing and cutting but is excellent for turnery, takes a good polish and finishes to a smooth surface. It is used for high-quality furniture, cabinets, decorative veneer, marine and aircraft grade plywood, ornamental turnery, carving, engraving, tool handles and sporting goods.

Dalbergia sissoo is reported to have pesticidal properties. Aqueous extracts from the leaves, stems and roots inhibit the reproduction, growth and development of the insect pest *Utethesia pulchella*. Methanol extract from the roots has insecticidal properties, especially against *Diacrisia obliqua*, *Spodoptera litura* and *Argina cubrania*. Oil obtained from the sissoo seeds is used to cure skin diseases. The powdered wood, applied externally as a paste, is reportedly used to treat leprosy and skin diseases.

Young branches and foliage form an excellent fodder with a dry-matter content of 32.5%. The foliage has normally been used as emergency feed when other fodder sources fail. A useful source of honey but the flowers are only lightly attached to the flower branch and fall easily. The bees are therefore not able to take full advantage of the large number of flowers. The species is fast growing, hence suitable for firewood. Sapwood and heartwood have calorific values of 4900 and 5200 kcal/kg, respectively.

The species is suitable for many agroforestry systems like agrisilvicultural, silvipastoral, hortisilvicultural systems. It may be planted as one component of a multitiered homegarden system, where it contributes several products. It is also used as a windbreak in mango, coffee and tea plantations. These shade-tolerant crops also benefit from improved soil fertility under *Dalbergia sissoo*. The tree nodulates and thus improves soil fertility.

KHAIRA (*ACACIA CATECHU*)

Acacia catechu is a small or medium-sized, deciduous tree up to 15 m tall with a light feathery crown and dark brown, glabrous, slender, thorny, shining branchlets, usually crooked. It is especially common in the drier regions on sandy soils of riverbanks and watersheds. It is native to India, Myanmar, Nepal, Pakistan and Thailand. In India, three varieties, namely var. catechu, var. catechuoides and var. sundra are recognized.

Propagation and nursery

Acacia catechu can be raised from direct sowing, coppice, planting out nursery-raised seedlings or by stump planting. The seeds adhere to the light pod valves after the pods dehisce and are often blown to a considerable distance from the trees. Seed fall takes place in the month of January and February. The seeds can be collected by lopping small pod bearing branches in January and spreading them in the sun for a few days. The seeds are separated from pods by thrashing. Seed viability is lost within 1 year at room temperature at 11–15% moisture content. There are 15,000–40,000 seeds/kg. Seeds are sown in the nursery in the month of April or May. It is recommended, but not necessary, to put the seeds in boiling water and then leave for 24 hours them to cool. Germination commences from about the 4th day after sowing and its completion may linger on up to 4–5 weeks.

Irrigation is essential in the nursery till the outbreak of monsoon. The seedlings require daily irrigation with a precaution that the water does not accumulate at the roots of the plants. The nursery should be kept free of weeds as these are liable to kill seedlings by suppression. Khair seedlings are comparatively resistant to damping off disease in the nurseries, however, water logging may sometimes predispose the seedlings to damping off in the early stage of development.

Direct sowing gives good results and it is very easy. Under optimum conditions, khair can also be propagated by stumps. The stumps should be made from seedlings about 15 months old raised in nurseries from seed sown in April of the previous year. Cuttings should be made from well developed seedlings. The root and shoot should be 23–31 cm and 2.5–5.0 cm, respectively. The best size of stumps at the root collar is 10–15 mm in diameter. Planting of stumps should be done soon after the break of rains.

Planting and tree management

Acacia catechu plantations are established in block or strip plantations at 1.5 m × 1.5 m to 3 m × 3 m. Since early growth is slow, weeding is essential, especially when the plants are still young. Protection from grazing by animals is required. Early thinnings are very important for the proper development of the crop. All shade, even lateral, must be removed. The first thinning is done at the age of 3 years. Subsequently thinnings are done at the ages of 5, 10, 15, 20 and 25 years. In coppice crops, it becomes necessary to reduce the number of the several coppice shoots sprouting from a single stump to one or two within 3–5 years.

Rotation regimes depend upon the intended use, for fuelwood production, felling is usually at 10–15 years of age. Trunks with a diameter of 30–35 cm are considered the most economic. For extracting the tanning agent cutch, this size is achieved only after 30 years.

Protection

The fungus *Ganoderma lucidum* causes serious mortality due to root rot in reforested stands. Khair is susceptible to the attack of pathogen at all ages. Young plants are killed soon after infection while the mature trees die when most of the roots become affected. The disease can effectively be checked by extraction of old stumps and cleaning of debris from the site, digging of isolation trenches in young plantations, planting of resistant species like *Bombax ceiba* and *Ailanthus excelsa* and mixed cropping (50:50) with resistant species.

Fomes badius causes heart rot in khair and is common in all khair forests, both natural and planted. The fungus causes decay in the heartwood only. Though sapwood remains healthy and free from infection the heart rot increased with age of tree and mature trees become unfit for extraction of cutch and katha due to its complete disintegration. The disease can be managed to some extent by avoiding injuries to the trees.

Uses

A substance called catechu (chemically, catechutannic acid), which is marketed as a solid extract, can be isolated from the heartwood khair. Depending on the way of processing, several products can be obtained from crude catechu. Dark catechu or cutch, which is mainly obtained as a by-product of the catechu industry is used for dyeing cotton and silk and preserving of fishing nets, sailing ropes and mail bags. About 63,000 tonnes of khair wood in India is annually consumed for manufacture of cutch and catechu.

Khersal, a crystalline form of cutch sometimes found deposited in cavities of the wood is used for the treatment of coughs and sore throat. The bark is said to be effective against dysentery, diarrhoea and in healing of wounds. The seeds have been reported to have an antibacterial action.

Though khair is chiefly used as a source of cutch and catechu, it is also a useful timber. The timber is comparatively heavy with a density of 880–1000 kg/cubic m at 15% moisture content. The wood is also very strong, durable and resistant to white ants. Timber is used for house posts, agricultural implements and wheels. The wood is excellent firewood. The calorific value of sapwood and heartwood is estimated to be 5142 and 5244 kcal/kg, respectively. Dry wood on destruction gives 38.1% charcoal of very good quality.

It is considered to be a good fodder tree and is extensively lopped to feed goats and at times cattle also. For leaf fodder, finger-thick branches are lopped usually before main leaf fall occurs. The leaves contain 13.0–18.7% crude protein.

EUCALYPTUS (*EUCALYPTUS* SPP.)

Eucalyptus is a genus of over seven hundred species of flowering trees. Most species of *Eucalyptus* are native to Australia. A few species are native to other countries. *Eucalyptus* have been grown in plantations in many countries because they are very fast growing and have valuable timber and can be used for pulpwood, for honey production or essential oils. In India *eucalyptus* is second most widely planted species after teak. Mainly five species of eucalypts are grown in different agroclimatic zones of the country. While *Eucalyptus camaldulensis, E. globulus* and *E. grandis* have restricted adaptability, *E. tereticornis* has been planted in the country both in the forests and outside in the agricultural fields on a large scale, except in the north-eastern states. Performance of species varies considerably with the site and climatic conditions. The suitability of the species in different agroforestry zones has been given in the following table. Trees are tall up to 50 m high. Bole is straight and clean with whitish mottled bark.

Suitability of eucalypts species in different agroclimatic zones of India

Species	Climatic zone	State
Eucalyptus camaldulensis	Tropical and sub-tropical	Himachal Pradesh, Uttrakhand
Eucalyptus globulus	Mild temperate to cool tropical climate	Kerala, Karnataka
Eucalyptus grandis	Lower subtropical areas	Kerala
Eucalyptus teriticornis	Dry tropical to moist tropical areas	Punjab, Haryana, Uttar Pradesh, Rajasthan, Gujarat, Madhya Pradesh, Maharashtra, Bihar, Odisha, West Bengal, Andhra Pradesh, Telengana, Karnataka and Tamil Nadu
Eucalyptus citriodora	Mild temperate areas	Hilly areas

Propagation and nursery

Natural regeneration is very poor due to destruction of seeds by ants. *Eucalyptus* leaves also contain germination inhibitor. Seeds should be collected from mature trees of 10 years or more old before the capsules dehisce. In a year seeds may be collected twice, February-March and October-November. Capsules should be dried in shade for one day and then gently shaken to clean the shells. Seeds are stored in tins at cool dry places. Seeds are viable up to three months. About 3,67,400 seeds weigh one kg.

Pretreatment of seeds is not required. The optimum temperature for germination is 32° C, but a wide range is tolerated. Seed should be sown in raised nursery bed in the month of October-November or February-March. Nursery soil should be sterilized with Aldrex or BHC against termites. Seed should be sown 20 g/m^2 of bed in lines 10 cm apart and 2.5 mm deep. The germination rate is generally high and can reach almost 100%. The seeds germinate in 4–15 days. 60–75 seedlings are obtained from one gram of seeds. Seedling can also be raised in seed trays. Trays should be kept moist with a fine spray of water until germination begins. The seedlings should be pricked out when they have two pairs of leaves into poly-bags of size 22 cm × 10 cm or root trainers. Polythene bags initially should be placed in shade for few days and then shifted in the open in sunken beds. The seedlings are hand watered for some time and later the beds may be irrigated. Three-month-old containerized stock is recommended for planting. *E. camaldulensis* is also suited to mass vegetative propagation. Cuttings from juvenile shoots root readily. Eucalyptus is a very good coppicer.

Planting and tree management

Though planting is done at the break of monsoon, eucalyptus can be planted throughout the year, provided irrigation is available. Planting is usually done in 20 cm × 20 cm × 20 cm pits at 3 m × 3 m spacing. Poor competition ability with weeds and the development of an open crown necessitate frequent weeding, until the canopy closes 3–5 years after planting. Depending on site conditions, *E. grandis* and *E. tereticornis* may respond to mineral fertilization with accelerated growth. Fertilizers may be applied at the rate of 30 g N, 30 g P$_2$O$_5$ and 15 g K$_2$O per sapling per year during the second, third and fourth years. For production of pulpwood and fuelwood, 6-10 year rotations are used without thinning. A thinning of 50% plants at 5 years provides posts, poles, fuelwood and pulpwood, leaving the better trees for the production of other products, such as sawn timber after 15 years.

Protection

Polyphagous insects seem to attack the nursery stock. Quinalphos or malathion 0.05% is recommended against these insects. Drenching the containers with chlorpyrifos is a preventive measure against termite attack in plantations. Quinalphos 0.2% solution is recommended to control stem borer attack.

Cylindrocladium leaf blight and pink diseases are common in eucalyptus trees. To control *Cylindrocladium* leaf blight, the nursery is drenched with carbendazim 0.05%. Bordeaux paste is recommended against pink disease. Using disease tolerant clones is a sure means of preventing the incidence of both diseases.

Uses

Eucalyptus grandis wood is pink to pale reddish brown in colour. It has good bending properties. It is used for house construction, floors, furniture, crates and veneers. *E. tereticornis* produces dark red wood. It is hard, strong, tough, heavy, very durable and resistant to termite attack. It is used for a wide range of construction applications. The timber requires special care in sawing and drying because of high incidence of spiral grain and has a tendency to split. *Eucalyptus globulus* is good for papermaking and is widely used for pulp. The density of the wood is 900–980 kg/m^3 at 12% moisture content. The firewood is suitable for industrial use in brick kilns but is not preferred for domestic use because it is too smoky and burns too fast. However, it makes good-quality charcoal.

E. camaldulensis is a major source of honey, producing heavy yields of nectar in good seasons. The honey is light gold and of reasonable density with a distinctive flavour. The oils extracted from eucalyptus leaves are used as an inhalant with steam and other preparations for relief of colds and influenza symptoms. Because of its refreshing odour and its efficiency in killing bacteria, the oil is also used as an antiseptic. It helps to treat lung infections, gastrointestinal ulcers and angina. The leaves are also used for the extraction of eucalyptol, a commercially important eucalyptus oil.

With its light crown, *Eucalyptus* is well suited for agroforestry systems. It is a valuable tree for windbreaks and shelterbelts.

CASUARINA (*CASUARINA EQUISETIFOLIA*)

Casuarina equisetifolia is a large evergreen tree with a straight bole and numerous, long, slender, drooping, jointed, leafless branchlets arising from rough woody branches. The jointed branchlets, which are partly deciduous, are green and perform the functions of leaves. Leaves are minute scale like and arranged in the form of a cup at the joints of the branchlets. Bark is brown, rough, fibrous and exfoliating in longitudinal strips. Although the tree is evergreen, it usually sheds a large amount of twigs throughout the year. Casuarina is native to Australia, the Indian subcontinent, south east Asia, and islands of the western Pacific ocean. The tree is widely planted throughout the tropics, especially for being able to provide timber whilst growing on poor sandy soils near the coast and thus also providing shelter and protection for the soil. The tree is commonly grown as an ornamental, especially near the sea where it can also offer shelter from the wind. In cultivation, *Casuarina equisetifolia* hybridizes with *Casuarina glauca* and *Casuarina junghuhniana*.

Propagation and nursery

Propagation is by seeds or through vegetative means. The fruit is a woody, oval structure superficially resembling a conifer cone, made up of numerous carpels, each containing a single seed. Seed viability is often low. For seedling production, about half kg seeds are sown on raised nursery beds of 10 m × 1 m. This will produce about 10,000 good quality seedlings. In sandy soils, farmyard manure is mixed with the topsoil. After sowing the seeds, a thin layer of sand is sprinkled to cover the seeds. Usually sowing is done in November-December. The nursery beds are kept moistened and proper shading is provided to facilitate rapid seed germination. Germination takes about 10 days and seedlings attain a height of 10–15 cm in 6 weeks. They are then pricked out into polythene bags or transplanted into beds of size 1 m × 10 m in January-February.

Vegetative propagation is done by branch cuttings, stump cuttings and layering. For vegetative propagation by rooting of branch cuttings, 5–7 cm long cladode cuttings are treated with rooting hormones. The hormone-treated cladodes are transferred to pre-soaked vermiculite and kept in a mist chamber. About 100% rooting is obtained within 15 days. The rooted cuttings are then transferred to a mixture of sand, soil and farm yard manure (2:1:1) for hardening. After 15 days, the hardened propagules can be transferred to the field. Air-layering is sometimes practised but is too costly for large-scale operations.

Planting and tree management

Casuarina has a wide environmental adaptability and hence occupies sites ranging from arid regions to coastal zones. Root nodules containing the actinorhizal symbiont *Frankia* enable *Casuarina equisetifolia* for biological nitrogen fixation. Therefore, it thrives best on sandy soils low in nitrogen and has the potential to improve the nitrogen capital of impoverished sites. It requires huge sunlight and a well-drained alkaline to neutral soil for proper growth. Plants can tolerate inundation by sea water at extremely high tides. Young plants are susceptible to drought until their roots reach the groundwater table, which may take up to 2–3 years after planting, but then they become very drought resistant.

Site preparation includes ploughing the land 2–3 times and making 30 cm × cm 30 × 30 cm pits before the onset of monsoon. The pits are filled with farm yard manure and topsoil. Planting is done immediately after the first rains. Casuarina can be planted in block, row and line or strips. Spacing varies depending on the objective and the end product. Usually a spacing of 75 cm × 75 cm is adopted. Young trees are susceptible to competition from weeds, especially grasses. Thus, one or two weedings may be required immediately after the rains. Timely thinning is essential as *Casuarina* species trees demand light. When the trees are about 3 m in height, the lateral branches are pruned to a height of about 2 m. Pruning is usually done at the end of the second year or after the beginning of the third year. In plantations established at close spacing, one thinning in the second year or third year depending on tree growth is desirable, where 25–50% of the trees are felled. In mixed species systems such as agroforestry, spacing and thinning practices are mainly dependent on the cropping systems and the nature of the associated species. If the associated crops are shade intolerant generally wider spacing and/or intensive thinning are recommended. Fertilizers may be applied at the rate of 20–25 g N, 15–20 g P_2O_5 and 15–20 g K_2O per seedling per year from the second year to the fifth year. The rotation period ranges from 4 - 5 years for fuelwood and 10–15 years for poles.

Protection

The common diseases in *Casuarina* nurseries are damping off, seedling blight, stem canker and seedling rot. Emisan 0.01% is effective against these diseases. Stem-wilt or bark blister disease caused by *Trichosporium vesiculosum* is a serious disease in the plantations. The disease affects trees of 3–4 years and causes mortality up to 80%. Maintaining a soil pH of 6.5–6.8 and treating the plantation with fungicidal sprays can control this disease. Other diseases include stem canker and dieback caused by *Phomopsis casuarinae*, pink disease caused by *Corticium salmonicolor*, root rot disease caused by *Ganoderma lucidum* and heart rot caused by *Polyporus glomeratus*, *Fomes fastuosus* and *Fomes senex*. Stem canker and dieback can be controlled by carbendazim @ 0.01%. There are no serious insect pest infestations inflicting extensive economic losses to casuarina trees.

Uses

Yield of high density fuelwood plantations varies from 10–20 tonnes per ha per year on 7–10 years rotations. Higher yields are reported from irrigated and fertilized sites. The highly regarded wood ignites readily even when green, and ashes retain heat for long periods. It has been called 'the best firewood in the world' and also produces high-quality charcoal. Calorific value of the wood is 5,000 kcal/kg and that of the charcoal exceeds 7,000 kcal/kg.

Wood is hard to very hard and strong. Heartwood is also resistant to termites. The wood is used for house posts, rafters, electric poles, tool handles, oars and wagon. The wood is used to produce paper pulp, and as a raw material for rayon fibres.

Casuarina plants are reported to have some medicinal values. Root extracts and barks are used for the treatment of dysentery, diarrhoea and stomach-ache. The powdered bark is used for treating pimples on the face. The cambium layer beneath the bark is squeezed and used to sedate a mentally ill or aggressive patient.

Since it is salt tolerant and grows in sand, the plant is used to control erosion along coastlines, estuaries, riverbanks and waterways. It is a nitrogen fixing species and therefore, often planted for reclaiming and improving the land. In the tropical cyclone-prone areas, casuarina is used as a protective planting due to its tolerance to strong winds. The abundance of highly branched twigs absorbs wind energy well. In areas with hot, dry winds the tree protects crops and animal herds. It is remarkably suited for boundary planting as it does not intercept much of the incoming solar radiation and yields substantial quantities of green leaf manure on lopping. With high productivity and properties that enhance soil fertility, it shows promise as an agroforestry species for arid and semi-arid areas. The tree is also grown as an ornamental, especially near the sea where it can also offer shelter from the wind.

TEAK (*TECTONA GRANDIS*)

Teak (*Tectona grandis*) is a large tropical tree reaching over 30 metres in height with an open crown that has many small branches. The bole, which can be unbranched for up to 15 m, is up to 1 metre in diameter. The species is indigenous to India and the Southeast Asian region. But it is naturalised and cultivated in many countries in Africa and the Caribbean. In India teak is distributed naturally in the peninsular region. It prefers a warm moist tropical climate with mean annual precipitation of 1100–2000 mm and a well-drained fertile soil. Being a strong light demander it does not tolerate overcrowding and does not withstand waterlogging.

Propagation and nursery

Generally stumps or seedlings are used as planting material. For stump preparation, the seedlings have to be maintained in the nursery for about one year.

Seeds should be collected from vigorously growing middle-aged trees characterized by straight boles, desirable branching habit, good form and less fluting. Freshly fallen intact fruits with inflated calyx from such trees are collected during December-February. Fruit is fleshy and bears 1-4 seeds which are enclosed in a stony covering. After cleaning and drying the seeds should be safely stored in gunny bags or sealed containers. Seeds of diameter greater than 9 mm are usually collected. For convenience in storage and transport, the bladder like calyx of the fruit is removed. This is done by half-filling a bag with the fruits and vigorously rubbing and shaking it or by beating with sticks, after which the remains of the calyces are separated from the nuts by winnowing.

Due to hard seed coat, germination of one-year old seeds is better than that of fresh seeds. Thus, it is necessary to break dormancy of the seed. Seed germination in the untreated seeds is totally absent or very in-significant. Several methods are in practice to break the dormancy of teak seeds.

- Seeds are spread over raised platform in August. The seeds get soaked in rains and then with sun they get dried thus getting natural weathering.
- Freshly collected seeds are put in gunny bags which are then submerged under flowing water for four days. The gunny bags are taken out and spread over in strong sun for four days. It is again submerged in water for 3–4 days following drying. It is repeated for 3–4 times until endocarp and mesocarp get easily cracked.
- Seeds are put in alternate layers of seeds and straw and daily watered for seven days then dried and stored till the time of sowing.
- Soaking the seed in concentrated H_2SO_4 for 20 minutes and thorough washing in running water hasten germination.
- Soaking the seeds in the mixture of cowdung and water enhance germination.
- Stored seed germinates best if soaked for 24–48 hours in warm water, prior to sowing, changing the water frequently.
- Another method of encouraging germination is to char (or half burn) the fruits by covering them with a thin layer of grass and lighting it.
- The freshly collected teak fruits are spread on the ground in a 5 cm layer immediately after collection. After about five weeks the termites remove the exocarp and subsequent germination after alternate wetting and drying is found to be better.

Raised seedbeds of about 30 cm high, 10 m length and 1 m width are prepared. A mixture of sand, soil and farmyard manure forms the top layer of the seedbed. Sowing is done after the bed is watered. Usually the sowing is done by broadcast method or dibbling in April-May. After sowing, the seeds are pressed into the beds and a thin layer of soil is sprinkled to cover the seeds. The beds are also mulched with green leaves to reduce evaporation losses. The bed is then dusted with carbaryl 10% to prevent insect attack. The germination rate is low, usually 60 - 70%. Germination starts after 10 days but may take 2–3 months. The seedlings can be transplanted to polythene bags after 3–4 months or it can be maintained in the nursery beds for 10–12 months for preparation of stumps. Seedlings should be given a little shade. One year old seedlings of 1–2 cm at the thickest portion below the collar are uprooted and all the leaves and secondary roots are removed. Stumps with 15–20 cm of taproot and 2–3 cm of stem are prepared with sharp knives. Stump planting is generally preferred and it is easy for transport. For seedling plantation, young seedlings are shifted to polythene bags containing soil mixture and maintained in the nursery for 3–6 months.

Planting and tree management

With the pre-monsoon showers, stump planting is done in crowbar holes. The site must be cleared of stubble or other competing vegetation. If seedlings are used, then optimal time of planting is after the onset of southwest monsoon. They are usually planted in pits of size 30 cm × 30 cm × 30 cm. Spacing recommended for monospecific woodlot is 2 m × 2 m. However, if intercrops are to be raised, then row-to-row distance can be altered. For one or two row strip plantings at farm boundaries, a closer plant-to-plant spacing of 1 m is recommended.

Teak is very susceptible to weed competition and frequent weedings may be required during the first two years. Fertilizers may be applied @ 30–40 g N, 15–20 g P_2O_5 and 15–20 g K_2O per plant per year from the second year to the fifth year and thereafter once in three to four years for 10–12 years. In agroforestry situations, if the intercrops are fertilized, the quantities of chemical fertilizers applied to teak can be proportionately reduced or even skipped. Providing life-saving irrigation during the summer season favours teak growth.

For a fifty-year rotation, monospecific teak plantation with initial spacing of 2 m × 2 m, thinning may be carried out at 4, 8, 12, 18, 26 and 36 years after planting. Thinning in short rotation (25–30 years) high input plantations can be at 4, 8, 12 and 16 years. The thumb rule governing thinning is that trees should not be allowed to compete with each other for site resources, as intense competition may depress teak growth. Therefore, considering the site characteristics, tree growth rate and merchantability of the thinned out materials, a flexible thinning schedule can be adopted.

Teak can be integrated to various agroforestry systems like, agrisilvicultural (teak + casuarina with agricultural crops maize, cotton, turmeric, tomato and chilli), agrisilvihorticultural (teak + guava/mango with agricultural crops pineapple, turmeric, vegetables, maize and cotton), silvihorticultural (teak + guava/annona) and silvipastoral (teak as tree component and napier and guinea as pasture components). Fruit/spice/medicinal trees also can be successfully intercropped with teak throughout its growth. Inclusion of nitrogen fixing trees such as *Gliricidia* or *Leucaena* either in alternate rows or every third row improves teak growth. However, the nitrogen fixing trees should be pruned regularly so that these do not compete with teak for light.

Protection

Hyblaea puera and *Eutectona machaeralis* are considered to be the major pests in teak. These insects are known to occur on seedlings in nurseries and also in grown up trees in plantations. *H. puera* feeds on tender foliages during the early part of the growth season and *E. machaeralis* feeds on older foliage towards the end of the season. These pests can be controlled by spraying of the foliage with endosulfan 0.05–0.075% or quinalphos 25 EC 0.05% spray.

Vascular wilt disease (*Burkholderia solanacearum*) is noticed in nursery and young plantations. As preventive measures against this disease, proper drainage should be ensured and root injury is to be avoided. Leaf spot disease (*Phomopsis* sp. and *Colletotrichum gloeosporioides*) in nursery and young plantations can be controlled by mancozeb 0.05% or carbendazim 0.05% application. Against pink disease (*Corticium salmonicolor*) in young plants, Bordeaux paste is applied.

Uses

Teak is one of the most important timbers in the world. A rare combination of superior physical and mechanical properties makes it a paragon of timber. Teak is recognized as the best timber for the manufacture of door, window frames and shutters, wagon and carriage, furniture, cabinets, ships, agricultural implements, decorative flooring and wall panelling because of its moderate weight, appropriate strength, dimensional stability and durability, easy workability and finishing qualities and most appealing grain, texture, colour and figure.

Various parts of the tree, including the wood are credited with medicinal properties. A wood tar paste is made from the powdered wood by putting it into hot water. It promotes digestion, relieves headaches and tooth aches and reduces inflammations of the skin. An oil extracted from the seeds promotes hair growth and the oil extracted from the roots is used to treat eczema, ringworms and inflammation. Flowers are considered useful against a number of diseases such as biliousness, bronchitis and urinary discharges. Leaves are used in indigenous medicine and their extract indicates complete inhibition of pathogenic bacteria *Mycobacterium tuberculosis*, the causative agent of tuberculosis. The leaves also contain yellow and red dyes, which have been recommended for dyeing of silk, wool and cotton. The leaves are occasionally used as plates for dining purposes, for making cheap umbrellas and for thatching temporary huts in some places. The bark is regarded as an astringent and considered useful in bronchitis. Activated charcoal can be prepared from its saw dust.

It is reported that the fine dust produced in machining operations may cause irritation of the skin or bronchial asthma and rhinitis after inhalation. Thus, a dust extractor fan is recommended. The substance responsible for the allergic reaction is probably the naphthoquinone desoxylapachol.

NEEM (*AZADIRACHTA INDICA*)

Azadirachta indica, commonly known as neem, is a fast-growing medium to large sized tree that can reach a height of 15-20 metres. The branches are wide and spreading. The fairly dense crown is roundish and may reach a diameter of 20-25 metres. The trunk is relatively short, straight and may reach a girth of 1.5-3.5 m. It is native to the Indian subcontinent. It is typically grown in tropical and semi-tropical regions. The neem tree is noted for its drought resistance. Neem is one of a very few shade-giving trees that thrive in drought-prone areas. The bark is hard fissured or scaly and whitish-gray to reddish-brown. The sapwood is grayish-white and the heartwood reddish.

Propagation and nursery

Neem regenerates naturally by seeds, coppice and root suckers. Fruit ripening coincides with rainy season. The fallen fruits germinate within a fortnight giving a thicket of seedlings under mature tree. It is easily raised in the nursery and planted out as potted plants or seedlings. Seeds are collected in June-July. Seed do not require any scarification or pre-sowing treatment, however, de-pulping and cleaning the seed considerably improve germination percentage. Germination is commonly 75-85% for cleaned seeds sown within a week of collection. Seeds do not retain their viability for more than 2 months and should be sown within 2 weeks of collection. Germination begins from 4 to 10 days after sowing and is usually complete within 3 to 5 weeks.

Seeds may be sown either in June-July in seedling containers or in open beds at a spacing of 2.5 cm in lines 15 cm apart. Depth of sowing is to 2.5 cm. The seeds should be lightly covered with soil and watered. Nursery soils should be loose and freely drained. Shade is not necessary but may be useful under extremely hot conditions. Seedlings grown in open beds should be thinned to a 15 cm × 15 cm spacing when they are approximately 2 months old. Pricking out of seedlings or wilding is done at three leaves stage into polythene bags. Seed can also be sown directly in poly bags.

Planting and tree management

Seedling growth in the nursery is moderate and is greatly hindered by competition. Plantations can be established at the beginning of the rainy season by direct sowing, by planting container-grown seedlings, or by using stump plants. In India, direct sowing has been found to be the most economical and successful method provided that seeds are sown as soon as possible after collection. Seedlings raised in the nursery can be planted when they reach 7 to 10 cm in height and have a taproot that is usually about 15 cm long. On very dry sites, larger seedlings of at least 45 cm in height are often preferred. Stumps from 1 to 2-year-old seedlings prepared by cutting back the stem to 2.5 cm and the taproot to 25 cm are commonly planted in India. Planting should be done in July-August in the pits of 1 m × 1 m × 1 m dimension. Cleaning around the pits should be done at the time of planting. Neem also responds well to chemical and organic fertilizers.

Neem can also be propagated vegetatively by air-layering, root and shoot cuttings, grafting and tissue culture. Successful air-layering can be achieved by treatment with 0.1% indole-butyric acid or naphthalene acetic acid.

Neem is a light-demanding species, but it tolerates moderate shade during the early stages of growth. It is intolerant of grass competition during the seedling and sapling stages, requiring weeding to ensure its survival, particularly in dry areas. It can withstand drought better than excessive rainfall. It is intercropped with a variety of field crops, usually cotton, sesame, and pigeon pea, greengram for up to 4 or 5 years.

Protection

Neem has few serious pests, but several scale insects have been reported to infest it. *Aonidiella orientalis* and *Pulvinaria maxima* feed on sap of young branches and young stems. In India, *Psuedocercospora subsesessilis* is the most common fungus attacking the leaves. The bacterium *Pseudomonas azadirachtae* may damage leaves.

Uses

Almost every part of neem tree is useful from its roots, trunk, bark, leaves, flowers, fruits and seeds in day to day life. Neem timber grain is rough and does not polish well. The wood is used to make wardrobes, bookcases and closets, as well as packing cases because its insect repellent quality helps to protect the contents from insect damage. The main stem of the tree is also widely used to make posts for construction or fencing because the wood is termite resistant. Charcoal made from neem wood is of excellent quality and the wood has long been used as firewood. Its oil is burned in lamps throughout India.

Azadirachtin is the principal active compound of neem plant. Extracts can be made from leaves and other tissues, but the seeds contain the highest concentrations of the compound. In India, many neem-based pesticides are available. These act as insect repellents, inhibiting feeding, and disrupting insect growth, metamorphosis and reproduction. Azadirachtin affects egg production and hatching rates. It can inhibit moulting, preventing larvae from developing into pupae. Many foliage-feeding species avoid plants treated with neem compounds or cease eating after ingesting the neem. Azadirachtin as an antifeedant is effective on about 100 insect species. Thus, the extracts work especially well to protect plants from defoliation without affecting beneficial pollinating insects like honeybees. A traditional agricultural practice involves the production of 'neem tea'. The seeds are dried, crushed and soaked in water overnight to produce a liquid pesticide that can be applied directly to crops. Crushed seed kernels are also used as a dry pesticide application, especially to control stem borers on young plants. These homemade remedies are often very effective in repelling pests or acting on insects as a feeding deterrent. The active compounds break down quickly, so an application of 'neem tea' generally provides protection for only about one week. Neem extracts may have toxic effects on fish and other aquatic wildlife and on some beneficial insects. Therefore, care should be taken that any unused extracts are disposed of by exposing them to heat or sunlight to break down the active compounds.

Neem has proved effective against certain fungi that infect humans. Various parts of neem tree have antiseptic, diuretic and purgative actions, and are also used to treat boils, pimples, eye diseases, hepatitis, leprosy, rheumatism, scrofula, ringworm and ulcers. Leaf teas are used to treat malaria. People use the twigs as toothbrushes, and dentists find twigs effective in preventing periodontal disease. Neem oil has been used traditionally as a typical treatment for skin symptoms in both humans and livestock. The young twigs and flowers are occasionally consumed as vegetables.

The large crown of neem tree makes it an effective shade tree, planted widely as an avenue tree in towns and villages and along roads in many tropical countries. Tree bark contains 12–14% tannins. Neem oil is used for preparation of soaps, cosmetics, pharmaceuticals and other non-edible products. An exudate can be tapped from the trunk by wounding the bark. This high-protein material has potential as a food additive and is widely used in Southeast Asia as 'neem glue'.

Farmers in India use neem oilcake (the residue left after extracting oil from the seeds) as an organic manure and soil amendment. It is believed to enhance the efficiency of nitrogen fertilizers by reducing the rate of nitrification and inhibiting soil pests including nematodes, fungi, and insects. Leaves and small twigs are used as mulch and green manure.

TAMARIND (*TAMARINDUS INDICA*)

Tamarindus indica is a large beautiful fruiting tree, growing up to 15 m tall with a dense, spreading crown. The trunk is short and the lower branches are borne almost horizontal. Fruit, a pod is indehiscent, sub-cylindrical, 10–18 cm × 4 cm, straight or curved and rusty-brown. The shell of the pod is brittle and the seeds are embedded in a sticky edible pulp. Tamarind is believed to be indigenous to tropical Africa, probably introduced long back to India by the Arabs.

Propagation and nursery

Tamarind can be raised easily through direct sowing, planting out of entire plants or stump planting or poly bag container plants. Plant begins to yield seed when about 8-10 years of age. Ripe pods are collected in the month of March-April. Seed is separated from edible pulp by washing with water then dried and stored. Dry seeds retain viability for several years at ambient temperatures.

Seeds are soaked for 24 hours in warm water and sown in an irrigated nursery seedbed or containers in April-May. Germination starts in about a week and about 80% germination is achieved in about a month. Germination is best when seeds are covered by 1.5 cm sandy loam soil. The seedlings quickly develop a taproot and so should not be allowed to grow in a nursery seedbed for more than 4 months before being transplanted into containers. Seedlings attain plantable size by July-August when they are about 3–4 months old and attain a height of 30–40 cm. The pit size should be 30 cm × 30 cm × 30 cm for final planting. Tamarind can be propagated through sowing in lines or patches directly in the main field. Depth of sowing should be about 1.5 cm. The lines are spaced at 4–5 m apart and the seeds are sown 10 cm apart. About 20 kg of seed is required to sow one hectare area. It gives 80% survival against 60% in transplanting.

Planting and tree management

Outstanding mother trees are propagated asexually through soft terminal cuttings, grafting, budding, or air-layering. The cuttings are prepared with current year's shoots and it is collected with new leaves flush early in the morning in turgid condition. Cuttings are treated with 1000 ppm indole-butyric acid.

Growth is generally slow. Seedling height increases by about 60 cm annually. The juvenile phase lasts up to 4–5 years. Young trees are pruned to allow 3–5 well-spaced branches to develop into the main scaffold structure of the tree. After this, only maintenance pruning is required to remove dead or damaged wood. Trees generally require minimal care. The species is strong light demander but resistant to drought. A long, well-marked dry season is necessary for fruiting. Size-control measures include close spacing of about 100 trees/ha and pruning to rejuvenate the fruiting wood. Spacing should be at least 13 m × 13 m in a pure plantation. Spacing may be reduced with vegetatively propagated plants, as they usually do not attain the same size as seeded trees. Smaller trees are easier to harvest. The trees also respond to coppicing and pollarding. Trees commence bearing fruit at 7–10 years of age, with maximum yields being obtained from about 15 years onwards. Trees can continue yielding for 200 years.

Uses

Sapwood is light yellow, heartwood is dark purplish brown; very hard, durable and strong, and takes a fine polishes. The wood is used for carving. It is used for general carpentry, sugar mills, wheels, agricultural tools, boat planks, toys, panels and furniture. Tamarind tree provides good firewood with calorific value of 4850 kcal/kg. It also produces an excellent charcoal.

The chief value of this tree lies in its fruits which are used for various types of food preparations and sherbets. The fruit pulp, mixed with a little salt, is a favourite ingredient of the curries and chutneys popular throughout India. The pulp consists of 8-14% tartaric acid. Acidity is caused by the tartaric acid. The ripe fruit of the sweet type is usually eaten fresh, whereas the fruits of sour types are made into juice, jam, syrup and candy. The juice has a high content of vitamin B (thiamine and niacin) as well as a small amount of carotene and vitamin C. The flowers, leaves and seeds can be eaten and are used in preparation of a variety of dishes. Tamarind seeds are also edible after soaking in water and boiling to remove the seed coat. Flour from the seed may be made into cake and bread.

The bark may be used to relieve sores, ulcers, boils and rashes. It may also be administered as a decoction against asthma and amenorrhea. Leaf extracts exhibit anti-oxidant activity in the liver, and is a common ingredient in cardiac and blood sugar reducing medicines. A sweetened decoction of the leaves is good against throat infection, cough, fever, and even intestinal worms. Filtered hot juice of young leaves is used for conjunctivitis. The pulp may be used as a massage to treat rheumatism. Powdered seeds are used to cure dysentery and diarrhoea.

Flowers are a good source for honey production. Both leaves and bark are rich in tannin. The bark tannins can be used in ink or for fixing dyes. Leaves yield a red dye, which is used to give a yellow tint to cloth.

Tamarind is not very compatible with other plants because of its dense shade, broad spreading crown and allelopathic effects. Thus, it is not suitable for agroforestry systems. It makes an ideal avenue tree by virtue of its shade, ornamental flowers and longevity. The deep roots make it very resistant to storms and suitable for windbreaks.

PHALSA (*GREWIA* SPP.)

Genus *Grewia* has several species out of which *Grewia asiatica* (known as *phalsa* in India and *falsa* in Pakistan) and *Grewia optiva* (known as beul or dhaman) are the most important and cultivated species in India. *Grewia asiatica* is extensively cultivated in India and in some other areas of the subtropics and tropics for its edible fruit, which fetches good prices in local markets. The fruit is an edible drupe, 5–12 mm diameter, purple to black when ripe. It is a deciduous small tree growing up to 8 m tall. *Grewia optiva* is a very popular tree of the farmers of the sub-Himalayan tract for its fodder and fibres. A full grown tree is moderate sized with spreading crown, reaching a height up to 12 m with a clear bole of 3–4 m and a girth of about 80 cm. The fruit is a fleshy drupe, 2–4 lobed, olive green when immature and black when ripe, and edible. The phalsa is indigenous to India and Southeast Asia.

Propagation and nursery

Seeds are the usual means of propagation of *Grewia asiatica* and *Grewia optiva* and they germinate in 15-20 days. Seeds do not require any special pre-sowing treatment but soaking in water for 12 hours may be required as seed coat is hard. Ground-layers, treated with hormones, have been 50% successful and air-layers up to 85%. Cuttings are difficult to root. Only 20% of semi-hardwood cuttings from spring flush, treated with 1,000 ppm naphthaleneacetic acid (NAA), and planted in July may root and grow normally. Long taproot is formed during the first year which is longer than the shoot length

In *Grewia optiva* fruiting starts in June-July and fruits ripen in October-November. The fruits are not borne on the current year's shoot, tree lopped completely do not bear fruits. Therefore, trees reserved for seed production should either not be lopped at all or if necessary only partially. The fruits are rubbed and washed in water to separate out the seeds. Each fruit contains 2-4 seeds. The seeds have a hard testa and can be stored well for at least a year without any appreciable drop in vitality. The dibbling method of sowing with twice a day irrigation proved to be the best for germination. The seeds are sown in March-April, about 2 cm deep in lines 15 cm apart. The seedlings are spaced about 10 cm apart in lines. Watering is done regularly till germination is over. The seedlings attain a plantable height of 30 cm by July.

Planting and tree management

Seedlings uprooted from the nursery with balls of earth are wrapped in moist gunny bags and transported to the planting site safely. Planting is done in

30 cm × 30 cm × 30 cm pits at a spacing of 4 m × 4 m for block planting and 4-5 m for single row planting along the field bunds. For stump planting 15 month aged seedlings are used.

Phalsa grows in both tropical and subtropical climates but will tolerate other climates, except at high altitude; however, it does best in regions having distinct summer and winter seasons. In India, it grows well up to an elevation of 1,500 m. The temperature range for growth is reported to be 10–44 °C with the optimum between 23–34 °C. Phalsa is a strong light demander and requires complete overhead light. The plant tolerates short periods of light frosts. Established plants are drought resistant. As seedlings are suppressed by weeds, weeding may be required at regular intervals at early growth period.

When grown commercially, *Grewia asiatica* plants are cut back hard annually to encourage the growth of new shoots. Flowers develop only on the new shoots of the current growing season. The period between flowering and fruit maturity is 45–55 days. Fruit yields are improved if the plant is cut back to the ground each year, thus encouraging plenty of new growth upon which the fruit is produced. Phalsa begins to bear fruit after only 2 years. The annual yield is 3–5 kg/plant or 4.5–6 tonnes per hectare. Summer is the fruiting season. Only a few fruits in a cluster ripen at any one time, so continuous harvesting is necessary.

Protection

Leaf-cutting caterpillars attack the foliage at night. A blackish caterpillar causes galls on the growing shoots. Termites often damage the roots. In some areas, leaf spot is caused by *Cercospora grewiae*.

Uses

The fruits of *Grewia asiatica* are eaten fresh as dessert or are made into syrup, and extensively employed in the manufacture of soft drinks. The squash is prepared from the fruit pulp by mixing it with sugar and used as an astringent, stomachic and cooling agent. The juice ferments so readily that sodium benzoate must be added as a preservative. The fruits are a cooling tonic and cure inflammation, heart and blood disorders and fevers. The fruit is also good against throat trouble. The bark cures urinary troubles. An infusion of the bark is used in treatment for diarrhoea. The root bark is employed in treating rheumatism. The leaves are applied on skin eruptions and they are known to have antibiotic action.

The wood of *Grewia asiatica* is medium-weight to heavy hardwood with a density of 730–900 kg/cubic m at 15% moisture content. Heartwood is pale grey to pale brown, not sharply differentiated from the sapwood; grain interlocked; texture fine; wood with some silver grain. The wood seasons well, is moderately soft to moderately hard, tough and moderately strong; it works satisfactorily with hand and machine tools. Non-durable when exposed to the weather or in contact with the ground, but durable for interior use. Under cover, the heartwood is moderately resistant to dry-wood termites. Wood is generally used for small articles where toughness is required, such as tool handles, spades, shafts of golf sticks, shoulder poles for carrying small loads, pestles, bows, billiards cues and shingles.

The wood of *Grewia optiva* is white, heavy, hard, elastic, strong and fine-textured. It is used for oar-shafts, axe-handles, shoulder poles, cat frames, bows and several other purposes, where strength and elasticity is required. The wood is difficult to saw when green but also difficult to work by hand after seasoning. It is reported to be suitable for paper-making. The shoots obtained after annual prunings are used for making quite strong baskets that are used to transport fruits and vegetables. Though used as a fuelwood, not liked very much because of the foul foetid smell it emits on burning. The bark yields a fibre of inferior quality used for ropes. The mucilaginous extract of the bark obtained after pounding in water is used to clarify sugarcane juice during the preparation of jaggery.

MULBERRY (*MORUS ALBA*)

There are three species of mulberry of economic importance, namely white mulberry (*Morus alba*), black mulberry (*Morus nigra*) and red mulberry (*Morus rubra*). *Morus alba*, is a fast-growing small to medium-sized mulberry tree which grows to 10-20 m tall. It is generally a short-lived tree. It is native to northern China and is widely cultivated and naturalized in the United States, Mexico, Australia, Kyrgyzstan, Argentina, Turkey, Iran, India, Cambodia, Indonesia, Japan, Laos, Myanmar, Pakistan, Thailand, Vietnam, Zanzibar, etc. The white mulberry is widely cultivated to feed the silkworms employed in the commercial production of silk. Its berries are poisonous when unripe, but are otherwise edible.

Morus nigra is a deciduous tree, slender but with numerous branches, grows to 6–9 m in height. Black mulberry is thought to have originated in the mountainous areas of Mesopotamia and Persia and are now widespread throughout Afghanistan, Iraq, Iran, India, Pakistan, Syria, Lebanon, Jordan, Palestine and Turkey. It has long been cultivated for its edible fruit in Europe and eastern China. The black mulberry was imported into Britain in the 17th century to be used in the cultivation of silkworms. It was unsuccessful because silkworms' preference to the white mulberry.

Morus rubra is a small to medium-sized deciduous tree, growing to 10–15 m in height and up to 50 cm in diameter. It lives up to 125 years. Red mulberry is native to eastern and central North America. Though it is common in the United States, it is listed as an endangered species in Canada. It is mainly cultivated for its large, soft, juicy and sweet with a good flavour fruits used to make delicious confection, pastries and wine.

Propagation and nursery

Morus alba is propagated either by planting out nursery raised seedlings or through rooted branch cuttings. Nursery raised seedlings may be planted out either as entire plants or as stumps. The stumps give better results than the seedlings. Direct sowing does not produce good results. Mulberry plants flower in March-April and fruits ripen in May-June. Trees of about five years age start producing viable seed. Ripe fruits are collected from the trees and are heaped in the shade, rubbed and washed in water to separate out the seed which is dried in sun for a few days before storage. Seed stored in gunny bags lose vitality after one year's storage. Thus, it should be stored in sealed tins. The seed stratified in layers of fine dry sand or ash keeps well for over two years.

For germination, the seed requires moist and well drained soil. Light shade is favourable for germination and seedling establishment. The seedlings can establish under canopy of trees having light crowns. Sowing in the nursery is done in May-June, soon after seed collection. Sowing is done in lines about 20 cm apart. Soaking of the seed in cold water for about a week hastens and ensures uniform germination. Pre-sowing treatment of the seed with kerosene oil or camphor water is also recommended to protect it from being carried away by ants. Since seeds are very small in size usually these are sown by mixing with sand or ash to ensure uniform sowing. Germination commences in about a week and is completed in another 10 days.

Planting and tree management

White mulberry or *Morus alba* grows in areas with a subtropical or mild temperate climate. The plant's water requirement is high. The shade-tolerant tree is highly susceptible to drought. It can withstand light frost. *Morus alba* grows on a variety of soils ranging from sandy loam to clayey loam. The tree cannot tolerate alkalinity and grows best on soils with pH ranging between 6.0 and 7.5.

The seedlings are transplanted in main site when about 10 cm tall at a spacing of about 60 cm × 60 cm. Transplanting is done in winter. All but a few terminal leaves are stripped before seedlings are planted in the main field. For the production of stumps, the seedlings may be retained in the nursery for one or two years depending on their growth rate. About 2 cm collar diameter is considered to be the most suitable size for stumps. Stumps with about 22 cm root and 8 cm shoot are prepared with a sharp tool so that these do not spilt.

These are wrapped in moist gunny bags during transport. Planting is done either in crow bar holes or in 30 cm × 30 cm × 30 cm pits. The spacing depends upon the objectives of raising the plantations. Close spacing may suffice if the trees are to be pollarded for leaf production. Wider spacing of 4 m × 4 m or 5 m × 5 m may be necessary if timber and leaf production are to be combined. The field is to be weeded when plants are young. Prunning may be required to regularize the shape of the plants and allow the growth of new shoots. The plants respond very well to coppicing and pollarding.

Uses

Morus alba yields a medium-weight hardwood. Heartwood texture is moderately with straight and attractive silver grain. It is easy to saw work, turn, bend and finish, and it seasons well. Wood is in chief demand for sports industry especially used for hockey sticks, tennis and badminton rackets, cricket bats, etc. It is also suitable for house building, boats, beams, posts, flooring, bridge building, agricultural implements, cabinet work, furniture, poles, shafts and bent parts of carriages and carts. It makes medium quality fuelwood with a calorific value of 4370-4770 kcal/kg. Wood yields sulphate pulp with satisfactory strength for white writing and printing paper.

The leaves are used for silk worm rearing. Leaves are also used as very good quality fodder for livestock. Dairy cows can be fed up to 6 kg of leaves a day to improve milk yield. Shade-dried leaves incorporated into feed enhance health and egg production in poultry.

Bark is said to be good in the treatment of stomach-ache, neuralgic pains and dropsy. The leaves and young branchlets are used for treating heavy colds, cough, red eye, insect bites and wounds, and the fruits are used in the treatment of sore throat, dyspepsia and melancholia. Fruit juice may be fermented and used to make liquor.

POPLAR (*POPULUS DELTOIDES*)

Poplars are amongst the fastest growing tree species under appropriate agroclimatic conditions. Genus *Populus* has 25–35 species, mostly native to North America. Few species like *Populus ciliata, P. laurifolia, P. gamblei, P. alba* and *P. glauca* are indigenous to the Himalayan region of India. However, an exotic species *Populus deltoides*, indigenous to United States of America became very popular in the north-western plains of Uttar Pradesh, Haryana and Punjab in the last few decades. Certain clones of *Populus deltoides* have been found to be eminently suitable for afforestation and agroforestry plantation. *Populus deltoides* is a fast growing tall tree with a fairly straight and slim trunk, open crown composed of a few large branches and attaining a height of 30 m and girth of 2 m. The branches are more or less angled or almost winged. Outer bark forms early, furrowed by cork-like ridges and deep fissures.

Populus deltoides can survive on soils varying from sandy loam to fairly stiff clay, but it makes its best growth on moist, well-drained, deep, medium-textured, alluvial soils that are fertile and well-aerated. Coarse sands and heavy clayey soils deficient in organic matter are unsuitable. It tolerates frost and waterlogging. *Populus deltoides* clones have been cultivated in a number of countries. In fact, *Populus deltoides* and its hybrids together make about 90% of the total cultivated poplars of the world. In India, it has been successfully cultivated as a forest crop or agroforestry crop in the Punjab plains and in the Terai region of Uttar Pradesh.

Propagation and nursery

The tree produces seed at intervals. Poplar seeds are very minute and 14,000 seeds weigh one kg. The seeds remain viable for about three weeks and should be used for sowing soon after collection. But it can be stored for a long period after properly drying. The viability of seeds can be maintained for several years by storing in cold storage at seed moisture content of 4-5% in sealed containers. Germination of *Populus deltoides* seeds is less than 20% under laboratory conditions. Thus, the plant is rarely raised by seeds due to its low germination rate. The seed bed is prepared with two parts soil and one part river sand duly sterilized to give higher survival of seedlings. Nursery beds are often fumigated with methyl bromide to control damping off. Seeds should not be covered or pressed into the soil of the seedbed. Young seedlings are very susceptible to drying out, and the seedbed must be kept water saturated for germination and at least one month thereafter.

Poplar is commercially propagated vegetatively through rooted cuttings and stumps. Cuttings should be collected from one year old shoots from lower $2/3^{rd}$ portion of the shoots or one year old plants from existing nurseries for better results. Cuttings from the main stem of the *Populus deltoides* give better results than those obtained from branches. Diseased, dying off-type and suppressed plants must be culled from the beds where stock is reserved for preparation of cuttings for the new nurseries. The plants can be cut at 2 cm from ground level. The length of the cuttings should be around 20–22 cm with thickness of about 1–2 cm having at least four buds. The upper cut of the cutting should be slightly above from the active bud. Cuttings should be prepared with a very sharp and fairly heavy tool to obtain a very clean and smooth cut and care should be taken to prevent splitting of cuttings. All cuttings must be submerged under fresh water in drums for 24 hours immediately after preparation of cuttings. Cuttings obtained from young, healthy and vigorous plants perform better. When cuttings are taken from the trees, the age of trees and of the parts of trees from which the cuttings are taken markedly affect rooting.

Cuttings treated with fungicidal solution for 10 minutes should be planted in nursery beds at a spacing of 80 cm × 60 cm by inserting them into holes made in beds. The best time of planting cuttings in the nursery is January to February. Cuttings made earlier than January and later than March do not perform well in open conditions. Each cutting with its thinner end up should be planted in the planting hole in such a way that the upper portion is just 2 cm above the ground level. After planting, the soil around each cutting should be compacted. The soil all around the cuttings should be drenched with Aldrex emulsion 0.2% immediately after planting the cuttings. Irrigation should be provided as soon as the planting of cutting in the bed is completed. The first irrigation should have about 5-7 cm water uniformly above ground level. Soil moisture in the nursery bed should be kept high during rooting of cuttings. Subsequent irrigation should be light and the interval may vary between 8 to 12 days depending upon the type of soil. The nurseries should be irrigated till the onset of monsoons and the topsoil should not be allowed to develop cracks and become absolutely dry. Proper and effective drainage of excess water during rainy season is essential to prevent water logging and collar rot. After the rainy season one to two irrigations per month is adequate.

As the poplar plants grow very fast, the nursery soil has to be enriched frequently. Application of farm yard manure @ 20-25 t/ha nursery area should be applied during bed preparation. Nursery beds are depleted of fertility after producing plants for one year, if no fertilizer is applied. After onset of the monsoon, urea @ 5 g/plant should be applied. Phosphorous application to nursery has also a good effect on root development. Weeding and hoeing may be required depending upon the incidence of weeds. Regular debudding is done by gently rubbing the newly formed buds with gunny bags up to 2/3rd height of plant from the base from June to October. But care must be taken not to damage the young leaves.

The nursery raised plants called 'entire transplants' (ETPs) attain a height of about 3 to 4 m in one growing season. These are utilized for planting in the main field. While removing these plants, the root is cut at 25 cm depth and all side roots more than 10 cm long are also trimmed. The plants are taken out with naked root and without foliage. ETPs can be prepared from one or two year old plants raised from cuttings in the nursery. This is the best method for transplanting poplars. Poplar propagation can also be made by stumps prepared from one year old nursery plants keeping 25 cm long root and 3 cm of shoot.

A lot of good quality clones of *Populus deltoides* have been developed and launched commercially. Some of them are G-3, G-48, D-61, D-67, D-121, L-34, L-49, L-52, WS-39, WSL-22, WSL-27, WSL-32, PL-1, PL-2, PL-3, PL-4, PL-5, PL-6, PL-7, etc.

Planting and tree management

Selection of proper sites for poplar plantations is of utmost importance because survival and growth rates of plants depend largely on the site quality. The land should be well drained as poplars do not withstand water logging. Assured irrigation facilities should also be ensured. Fertile loam or silt loam soils rich in organic matter should be preferred for poplar plantation. Areas affected with salinity or alkalinity should be avoided.

One deep ploughing up to 25-30 cm depth is desirable before final planting of poplars. A light preparatory irrigation may be required to bring the soil to proper moisture to facilitate ploughing. The field should be levelled after ploughing. Phosphatic and potassium fertilizers should be added during land preparation. As poplars are very sensitive to zinc deficiency, zinc sulphate @ 25 kg/ha should be applied in the area at the time of site preparation.

The best time for planting poplar is during the months of January-February before the opening of new buds. One year old ETPs of 3 m length containing 25 cm root portion give the best results. Nursery beds should be given a light irrigation 7-10 days before uprooting so that there will be no damage to the roots or any injury to the stem while uprooting. Bundles of ETPs should be placed horizontally in the pits filled with fresh water to be replenished from time to time. Storage under fresh water for minimum 48 hours is desirable. If the ETPs suffer dehydration and damage to phloem and cambium layers the establishment and survival of the plants will be difficult.

ETP should be held vertically in the planting pit which should then be filled completely with soil. In block plantation a spacing of 4-5 m × 2 m is maintained. The best spacing under agroforestry systems is 5 m × 4 m. The depth of planting will depend upon the soil type, depth of water table, size of planting stock, etc. As poplar is a fast growing species, the crop should be adequately fertilized for proper growth. A basal dose of 2 kg farm yard manure, 50 g single super phosphate and 25 g muriate of potash per plant are applied. Nitrogenous fertilizer should be applied in split doses, first dose of nitrogen as 75 g urea should be applied during the first week of June, second dose of 150 g urea during first week of July and third dose of 250 g urea during second and third week of August. Application of fertilizer must be followed by light irrigation.

Sprouting starts by the end of March and is completed by end of April. Certain clones sprout fairly late up to the middle of May. Some of the plants may not sprout at all and this can be ascertained by inspecting the buds of the plants. If the buds do not show signs of swelling, the plants should be cut back at the ground level in early May. Special care by way of irrigation and soil working should be taken for the cut back plants which will sprout if the same were green and alive at collar level at the time of cutting back. Precautions should be taken to ensure that wet or very moist soil is not heaped around the stem of the plants as that may encourage collar rot.

Timely irrigation holds the key to the successful establishment and growth of poplar plants. As the poplars grow and root system develops and spreads over large parts of the plantation area flood irrigation during summer months in beds to cover the entire area will be very helpful. However, flooding may result in lodging if there is a strong wind storm.

Pruning is essential for ensuring a large clear bole with concentration of maximum volume in the main stem for improving yield and quality of commercial timber. During the first year of growth debudding operations should be carried out in the lowest one third part of the stem during June-July. Pruning of branches at the lower portion of the stem should be carried out as per requirements. However, between second and third years of age pruning should be restricted to a maximum of lower one-third part of the total height of the tree. During subsequent years pruning could be carried out up to a maximum of half of the total height of the tree. Main stem should not be damaged during pruning. Bordeaux paste should be applied immediately after carrying out pruning. After one year pruning should be carried out only during winter and not in summer or rainy season.

Protection

Major insect pests causing damage to the poplars are leaf defoliator, stem and shoot borer, leaf webber, bark eating caterpillar, case worm and leaf hopper. Leaf defoliators are active during the rainy season while leaf webbers are active from April to November with peak period from July to October. Chlorpyriphos 500 ml + cypermethrin 200 ml/ha should be sprayed to control these pests. Stem and shoot borer can be controlled by applying phorate 10 G @ 12 kg/ha up to second year of plantation. After 2 years, these can be controlled by pushing a small wick of cotton dipped in any liquid fumigant in the holes through which frass is being pushed out by the borer. All holes must be closed with mud paste after such treatment. Bark eating caterpillar and case worm can be controlled by spray of 2.5 kg carbaryl 50 WP/ha at feeding sites and the infested branches should be heavily pruned. The leaf hoppers are active from April to November with peak period of their activity from July to October. A spray of oxydemeton-methyl 25 EC @ 750 ml or dimethoate 30 EC @ 625 ml/ha is recommended to control this pest. If mite infestation is observed metasystox @ 2 ml/l of water should be sprayed 2-3 times at an interval of 15 days. Termites can be controlled by applying chlorpyriphos @ 3.75 l/ha.

The cuttings are dipped in 0.5% solution of emisan-6 for 15 minutes before planting as preventive measure against collar rot. If blight infestation is observed the crop should be sprayed with carbendazim @ 2 g/l of water. Wilt can be controlled by spraying of wettable sulphur powder @ 1.5 kg/ha.

Poplars can be harvested at short rotations of 8 to 10 years. Poplars can attain 90 cm girth at breast height and mean annual increment of 20 m^3/ha at 8 years rotation under good care.

Uses

Populus deltoides has a medium hardwood, light, free from knots, easy to saw and work. The wood has good nail holding power and can be used for making cases for fruit and food stuffs. Wood obtained from poplars is suitable for manufacture of match splints, veneering products, artificial limbs, interior panelling, sports goods, and cheap furniture, etc. It is also suitable for structural uses such as false ceilings, partition and almirah shelves, etc. However, most of the poplar wood is used in the match and plywood industry. The species is suitable for making general purpose plywood, marine plywood, concrete shuttering plywood, etc. Young poplars of 2-3 years age group are an excellent source of cellulose fibre for making various grades of fine paper, packing paper, newsprint, etc. Wood is used for fuel.

Any traditional crops except paddy can be grown reasonably well in between the lines of poplars during the first 3 years. Sugarcane is preferred as an intercrop for the first two years as it is more profitable. Subsequently shade tolerant crops like ginger, turmeric, colocasia, pineapple, arrowroot, etc. can be raised successfully. Short duration winter vegetables or *rabi* crops like wheat, oat, potato, coriander, soybean, lentil, chickpea, toria, cabbage, tomato, etc. can also be raised as most poplar clones are leafless during autumn. Third year onwards cultivation of these crops can be raised throughout the rotation. However, intercrop yield decreases with the increase in age of poplars.

OAK (*QUERCUS* SPP.)

The oak tree is one of the most loved trees in the world. It is a symbol of strength, morale, resistance and knowledge. The oak tree belongs to the genus *Quercus* which has about 800 species all over the world, especially in the Northern hemisphere where they are native. It is a long-lived tree which can live more than 1000 years.

Oak is moderate to large sized tree with almost rounded crown. It attains a height of 20 m and diameter of 60 cm. It is an evergreen tree, old leaves fall after the appearance of new leaves and thus trees are never leafless. New leaves appear in the month of March-April.

Propagation and nursery

Oak can be propagated easily through direct sowing or planting out of nursery raised seedlings. Air layering is also possible and application of auxins encourage rooting. In oak, male catkins and female spikes appear on new shoots in April-May. The fruit takes 6-18 months to mature, depending on their species. The oak fruit is a nut called an acorn or oak nut borne in a cup-like structure known as a cupule. Each acorn usually contains one seed and rarely contains two or three. Acorns are collected in December-January preferably from tress and are dried under shade only. The seeds can be stored for one year at low temperature and high humidity.

Oak can be successfully raised by direct sowing either dibbling the seed in cultivated line or patches. The seeds are sown about 2 cm deep before the onset of the rains. About 5 kg of acorns are sufficient for one hectare area plantation. Oak seeds germinate in heavy or moderate shade but fail to germinate in places exposed to direct sun. Drought is the most adverse factor causing seedling mortality. Fresh seed gives 60–70% germination. Germination starts in about 10–12 days and takes to 4–5 days to complete. Seedling develops a long taproot and the growth of root is generally more than that of shoot during the first few years.

In case of nursery, beds should be prepared under light shade, and not under direct sunlight. Seeds are sown in lines about 20 cm apart with 5 cm spacing within the lines at a depth of 2 cm in raised beds. Deep sowing delays the germination and also reduces germination percentage.

Planting and tree management

Regular weeding is required in nursery beds. Seedlings soon develop a taproot and become intolerant of root disturbance. So they should be planted into their permanent positions while young. Seedlings of about 15–20 cm height are suitable for planting out as taller seedlings are difficult to handle due to their massive roots. Planting should preferably be completed by July. Seedlings should be planted at a spacing of 3–4 m × 2–3 m. Oak species are slow growing and thus these need weeding and cleaning for several years.

Oak is moderate light demander and can tolerate to shade in early age. But older trees benefit from full light. Seedlings are sensitive to frost and must be protected during the winter, but older trees are more frost-resistant.

Uses

The major use of oak is for fuelwood and charcoal making, for which it is in great demand. Calorific value of fuelwood is 4600 k cal/kg. Tree is lopped for fodder and a mature tree can yield 20–25 kg leaves annually. The wood is very hard, strong and durable. It has a good grain if properly cut, but does not season well, being rather apt to warp and shrink. The wood is used for light construction purposes, agricultural implements, axe-handles, poles, toys, etc.

Oak seeds, raw or cooked have a sweet taste. It can be eaten whole, though it is more commonly dried, then ground into a powder and used as a thickening in stews or mixed with cereals for making bread. The roasted seed of many *Quercus* species has been used as a coffee substitute. The oak seeds are also used in the treatment of diarrhoea, indigestion and asthma. The leaves of most species in this genus are more or less rich in tannins. The bark of oak trees is also usually rich in tannins.

BAMBOO (*DENDROCALAMUS STRICTUS*)

Bamboos are tall grasses consisting of large canopies. Different species of bamboo are found both naturally and cultivated in farms. *Dendrocalamus strictus* is the most common bamboo species which is widely distributed throughout India, Myanmar and Thailand. This is the hardiest of all Indian bamboos. *Bambusa bambos* is another species occurring throughout India, except in the Himalayas and sub-Himalayan region and the valleys of Ganges and Indus. It is very common in Odisha, the Western Ghats, and throughout south India. Besides these two dominant species, several other bamboo species and sub-species are widely found in India and other south-east Asian countries. These include *Bambusa balcooa, B. glaucescens, B. longispiculata, B. nutans, B. polymorpha, B. tulda, B. vulgaris, Dendrocalamus giganteus, D. hamiltonii*, etc. Different species have different growth rates, development patterns and different requirement of environmental conditions.

Dendrocalamus strictus also known as male bamboo, solid bamboo or Calcutta bamboo is a tropical and subtropical clumping species. It is native to India where it occupies 53% of the total bamboo area. *Dendrocalamus strictus* is a medium-sized bamboo with culms of about 8-20 m tall and 2.5-8 cm in diameter. The internodes are 30-45 cm long and thick-walled. Culms are hollow under humid conditions, but nearly solid under dry conditions. This species has pale blue-green culms when young, and dull green or yellow culms on maturity. Its nodes are somewhat swollen and basal nodes are often rooting.

Propagation

Dendrocalamus strictus prefers a low relative humidity and mean annual temperatures between 20°–30° C, but can withstand extreme temperatures, as low as –5° C and as high as +45° C. The optimum annual rainfall is 1,000-3,000 mm, with 300 mm per month during the growing season, but is very drought resistant. The species does not grow well on water-logged or heavy soils. It prefers sandy loam soils with good drainage and a pH 5.5–7.5.

Bamboos can be propagated through seed or by vegetative means. Bamboo seeds should be sown within 2–3 months of harvest. Seeds are sown in containers in a lightly shaded position. Germination usually takes place readily. Seedlings are pricked out into individual pots as soon as these are large enough to handle. Seedlings are then planted out into permanent positions when 20 cm tall.

However, some species such as *Bambusa balcooa* and *Bambusa vulgaris* do not produce seed and some other species such as *Bambusa bambos* and *Dendrocalamus strictus* do flower at long intervals varying from 30 to 70 years. Some species flower gregariously and then the parent plants all die, some flower sporadically with or without parent plant deaths, and some species combine both patterns. Only a few species flower more frequently and set seeds. Moreover, bamboo seeds are short-lived and difficult to store outside research institutions. Thus, bamboos are mostly propagated by vegetative means like rhizomes, culm cuttings and branches. However, for this high air humidity, appropriate rooting medium, moderate to full light intensity, protection from pests and diseases, waterlogging and strong winds are required. Of these, maintenance of high air humidity and lack of water logging are the most critical.

Rhizomes
Natural regeneration of bamboo occurs through rhizomes. The rhizome is an underground stem, closely similar in structure to the above ground portion of the stem. The buds on rhizomes, which usually develop are generally one year old. Separated out rhizomes can be planted in the rainy season. Basically there are three ways; use of rhizome with roots, rhizome with culm and roots or rhizome with culm-stock and roots.

1. *Rhizome with roots*: Rhizomes should be healthy not more than 2-3 years old; not damaged; and should have roots. Rhizomes without culms are cut 50–60 cm long with roots.
2. *Rhizome with culm and roots*: Generally young culms along with associated rhizomes and roots are used. If the culm is young, it is maintained intact, if it is large the upper part can be removed, or branches trimmed. It is essential that the rhizome attached bears buds; other rhizomes can be cut away.
3. *Rhizome with culm-stock and roots*: This is essentially a rhizome with roots but only 30-60 cm of the basal portion of the connected culm.

February to March appears to be the best time for both collecting and transplanting. If the region has a cold climate the best planting time is April, if it is warm then it should be the cooler month. If rhizomes alone are used then they need a period in the nursery during which they are laid horizontally to root.

The two other types are transplanted directly into the field. In a rooting bed the medium should be soil and sand in 3:1 ratio. The rhizome are planted 10-15 cm deep, and 25 cm apart. Watering is necessary and straw mulch is helpful. Shade is needed when young shoots emerge. If many plants are to be produced from one piece of rhizome, they are to be separated and usually 1-2 sprouts may be kept per piece. The piece should be at least 15-20 cm long, with three or four nodes, each with an intact bud. The length of the rhizome affects the survival of the shoot. The longer the rhizome, the more nutrient it contains to support the new shoot. Rhizomes are planted in pit of 60 cm^3 and at spacing of 6-8 m × 6-8 m.

Culm cuttings

The culm is the above ground stem which grows from the underground rhizomes. It is segmented with nodes and internodes. The length of internodes and their texture vary according to species. The nodes bear one or more buds, their shape and size also depend on species. The buds occur on alternate sides of the culm.

Culm cuttings are segments of the culm usually 1 or up to 2–3 nodes with buds or branches. They are suitable for clump-forming species but usually not for non-clump-forming species. Generally the culm selected for the cuttings should be not more than 2 years old and buds should be healthy. Segments are selected from the lower to mid zone of the culm. The upper part and the lateral branches of the upper culm are discarded. Mid-March to May is the best time for cuttings. The branches on the selected part of the culm are pruned to a length of 10–30 cm, care being taken not to injure existing buds.

A segment is cut with a sharp knife or saw. There must not be any splitting at the cut. A length of 5–10 cm stem should be kept on either side of the node. Immediately after cutting, the cut ends are waxed or segments are wrapped with moist gunny bags to minimise water loss from cut ends. If cuttings are 2 or more noded, an opening of 2 cm long and 1 cm wide is made, or 2 holes each 7 mm diameter are drilled in the centre of the internode. A rooting hormone solution is poured into the segment cavity and the holes are closed by wrapping and tying a polythene strip.

Branch cuttings

Propagation through branch cuttings is one of the most practical methods due to ease of handling. Like culms, branches are stem material. Thick-walled species with stout branches are ideal e.g., *Bambusa* spp. or *Dendrocalumus* spp., but survival sometimes is only about 50%. Branches require 6–12 months for rooting, and 12–30 months for rhizome development essential for production of new culms.

Branching patterns vary among bamboo species. Each bud or culm node can potentially produce a branch; however, in many species the primary branch remains dominant and stout. An offset is the lower part of a single culm (usually with 3–5 nodes i.e., about 1–2.5 m) with the rhizome axis basal to it and its roots. Planting of these is the most conventional way of propagating bamboo. The culm is between 1 and 2 years old. The culm is cut with a slanting cut and the rhizome to which it is attached is dug up and cut off to a suitable length to include well developed buds. Offsets are normally obtained just before the rainy season or after a pre-monsoon shower. Collecting of offsets is done 2–3 months before planting; in this case they should be kept in a temporary nursery near the site.

Propagation bed
Nursery beds are 1 m wide and range from 5 to 15 m in lengths depending on space available. They should be built on level ground and should be 20–30 cm deep. At the base, medium to coarse sand should be laid for 10–15 cm. Over that a top layer of fine sand should be laid for 10–15 cm. Before planting the cuttings, their branches are trimmed. Culm cuttings are placed horizontally in the bed with the bud upward, spaced 15–30 cm apart and so that branches if present are arranged away from those of another cutting. The cuttings are placed in the top fine sand layer about 3–5 cm above the coarse sand and covered by 3–6 cm of the fine sand. The bed is kept moist. Partial shading is required.

A new technology for mass production of bamboo seedling has been developed by the Forest Research Institute, Dehradun. Seeds are sown in July in the germination trays and when the seedlings reach 3–4 leave stage they are planted in polybags containing equal proportion of soil, sand and FYM. At the age of eight month, seedlings are removed from polybags. Proliferated tillers of these seedlings are separated by cutting rhizome to act as propagules. Each propagule consists of a tiller along with rhizome and roots.

Management of established clumps

Bamboos have an interesting method of growth. Each plant produces a number of new stems annually. These stems grow to their maximum height in their first year of growth, subsequent growth in the stem being limited to the production of new side branches and leaves. Bamboo responds well to fertilizer application. Organic fertilizers are recommended if grown for edible shoots. Irrigation may be essential during the first two years to ensure better establishment and quicker culm production. If grown for edible shoots watering ensures enhanced sprout production. Regular thinning and cleaning should be carried out from the 4th year of clump establishment.

All dry, dead and drying culms are to be removed from the clump so as to create sufficient space in the clump for new sprouts to grow up straight. As a regular practice these operations are to be carried out every year in the months of November-February. If managed properly with routine pruning, thinning and cleaning, bamboo usually escapes pest infestations. Proper sanitation measures should also be adopted for the control of fungal infections.

No current year culms are to be harvested. About 60% of the 3 year old culms and almost all of the 4th year culms can be cut and removed. However, it is always better to retain a few older culms in the clump to serve as support for the younger newly emerging culms. Harvesting should never be done during the growing season. It is recommended to cut the culms lower than 30 cm above the ground level, but not below the 2nd node.

If the stand is managed for edible shoots, they are to be cut either in the early morning or late evening when the sprouts attain 35-45 cm in height. Only 60% of the sprouts from all portions of the clump should be harvested while 40% are retained in the clump for culms production.

Uses

Dendrocalamus strictus is extensively used as raw material in paper mills and also for a variety of purposes such as light construction, furniture, musical instruments, bamboo board, mats, walking sticks, agricultural implements, rafts, baskets, woven wares and household utensils. Young shoots are edible and used as food. Leaves are used as forage. Many seasonal field crops like toria, sesame, pulses, vegetable, etc. can suitably be intercropped in agroforestry systems with bamboo.

Selected References

Baker, F. W. G. 1992. Rapid propagation of fast-growing woody species. CAB International, Wallingford, U. K.
Balasubramaniyan, P. and Palaniappan, SP. 2001. Principles and Practices of Agronomy. Agrobios India, Jodhpur.
Bandopadhyay.1997. A Text Book of Agroforestry with Application. Vikash Publishing House Pvt. Ltd., New Delhi.
Bene, J. G., Beall, H. W. and Cote, A. 1977. Trees, food and people: Land management in the tropics. IDRC, Ottawa, Canada.
Betters, D. R. 1988. Planning optimal economic strategies for agroforestry systems. *Agroforestry Systems* 7: 17-31.
Chundawat, B. S. and Gautam, S. K. 1993. Textbook of Agroforestry. Oxford and IBH Publishing Company Pvt. Ltd., New Delhi.
Dhyani, S. K. 2008. Current status of agroforestry research in India. *Abstracts*. National Symposium on Agroforestry knowledge for sustainability, climate moderation and challenges ahead. NRC for Agroforestry, Jhansi, India.
Dhyani, S. K. 2018. Agroforestry in Indian Perspective. In: Rajeshwar Rao G, Prabhakar M, Venkatesh G, Srinivas I and Sammi Reddy K (Eds.) Agroforestry Opportunities for Enhancing Resilience to Climate Change in Rainfed Areas, ICAR - Central Research Institute for Dryland Agriculture, Hyderabad, India. pp. 12-28.
Divya, M., Parthiban, K. T., Vanangamudi, K., Srinivasan and Govindarao, M. 2008. Social forestry and agroforestry. Satish Serial Publishing House, Delhi.
Dwivedi, A. P. 1992. Agroforestry Principles and Practices. Oxford and IBH Publishing Company Pvt. Ltd., New Delhi.
FAO. 1989. Arid zone forestry, a guide for field technicians. FAO Conservation Guide No. 20, Rome.
Gupta, J. P. and Sharma, B. M. (Eds.) 1997. Agroforestry for sustained productivity in arid region. Scientific Publishers, Jodhpur, India.
Gupta, R. K. 1993. Multipurpose trees for agroforestry and wasteland utilization. Oxford and IBH Publishing Company Pvt. Ltd., New Delhi.
Huxley, P. A. 1983. The Tree/crop Interface. ICRAF Working Paper No. 13. ICRAF, Nairobi, Kenya.
Huxley, P. A. 1984. The basis of selection, management and evaluation of multipurpose trees. ICRAF Working Paper No. 25. ICRAF, Nairobi, Kenya.
Jaenicke, H. 1999. Good Tree Nursery Practices: Practical Guidelines for Research Nurseries. International Centre for Reseach in Agroforestry, Nairobi, Kenya.
Jha, L. K. 1995. Advances in Agroforestry. APH Publishing Company, New Delhi.

KAU. 2002. Package of Practices Recommendations: Crops, 12th Edition. Kerala Agricultural University, Trissur, Kerala.

Lundgren, B. O. and Raintree, J. B. 1982. Sustained agroforestry. *In*: Nestel, B. (Ed.). Agricultural Research for Development: Potentials and Challenges in Asia. ISNAR, the Hague, the Netherlands.

Maji, A. K., Obi Reddy, G. P. and Sarkar, D. 2010. Degraded and Wastelands of India: Status and Spatial Distribution. Indian Council of Agricultural Research, New Delhi.

Nair, P. K. R. 1984. Soil Productivity Aspects of Agroforestry. ICRAF, Nairobi, Kenya.

Nair, P. K. R. 1987. Agroforestry systems in major ecological zones of the tropics and subtropics. ICRAF Working Paper No. 25. ICRAF, Nairobi.

Nair, P. K. R. 2008. An Introduction to Agroforestry. Springer (India) Pvt. Ltd., New Delhi.

NCA. 1976. Report of the National Commission on Agriculture: Part V, IX and Abridged. Ministry of Agriculture and Irrigation, Government of India, New Delhi.

Negi, S. S. 1986. A Hand Book of Social Forestry. International Book Distributers, Dehradun, India.

Negi, S. S. 1999. Agroforestry Handbook. International Book Distributers, Dehradun, India.

Ong, C. K. and Huxley, P. (Eds.). 1996. Tree-crop Interactions: A Physiological Approach. CAB International, Wallingford, U. K.

Pandey, D. N. 2007. Multifunctional agroforestry systems in India. *Current Science* **92**: 455-463.

Pathak, P. S., Pateria, H. M., Solanki, K. R. (Eds.). 2000. Agroforestry systems in India. NRCAF, Jhansi, India.

Pathak, P. S., Solanki, K. R., Rai, P., Handa, A. K. and Pateria, H. M. 2009. Agroforestry. *In*: Handbook of Agriculture, 6th Edition. Indian Council of Agricultural Research, New Delhi.

Patra, A. K. 2011. Agroforestry for food and wood. *Science Horizon* **1**(7): 24-27.

Patra, A. K. 2013 Agroforestry: Principles and Practices. New India Publishing Agency, New Delhi.

Patra, A. K. 2017. Agroforestry: A Sustainable Land-Use System for Food and Wood. *Everyman's Science* **55**(5): 290-296.

Patra, A. K. and Patra, A. R. 2018. Social Forestry: Why and Where. (*In*) 'Geo-spatial Technology and Ecosystem Assessment' (eds., C K Bharti, A Chauhan and P K Bharti). Discovery Publishing House Pvt. Ltd., New Delhi. pp. 147-169.

Puri, S. and Khosla, P. K. 1993. Nursery Technology for Agroforestry- Application in semiarid region. Oxford and IBH Publishing Company, New Delhi.

Rao, K. P. C., Verchot, L. V. and Larman, J. 2007. Adaptation to climate change through sustainable management and development of agroforestry systems. *SAT eJournal* **4**:1-30.

Shah, S. A. 1988. Forestry for people. India Council of Agricultural Research, New Delhi.

Tejwani, K. G. 1994. Agroforestry in India. Oxford and IBH Publishing Co. Pvt. Ltd., New Delhi.

Tejwani, K. G. 2008. Agroforestry to meet the challenges of the Indian agriculture. *Abstracts*. National Symposium on Agroforestry knowledge for sustainability, climate moderation and challenges ahead. NRC for Agroforestry, Jhansi, India.

Tejwani, K. G. Druvanarayan, V. V. and Satyanarayan, T. 1960. Control of gully erosion in ravine land of Gujarat. *Journal of Soil and Water Conservation* **11**:20-21.

Umrani, R. and Jain, C. K. 2010. Agroforestry systems and practices. Oxford Book Company, Jaipur.

Westoby, J. 1989. Introduction to World Forestry: People and their Trees. Basil Blackwell, Oxford, U. K.

Young, A. 2005. Agroforestry for soil management. CAB International, Walling Ford, U. K.

Web Links

http://en.wikipedia.org
http://forestry.about.com
http://www.agroforestry.co.uk
http://www.agroforestry.net
http://www.americanforests.org/discover-forests/forest-facts
http://www.bioversityinternational.org
http://www.centerforagroforestry.org/practices
http://www.conference.ifas.ufl.edu
http://www.earo.org.et/communication
http://www.ejournal.icrisat.org
http://www.eoearth.org/article/agroforestry
http://www.fao.org/forestry
http://www.greenbeltmovement.org
http://www.greenbooks.co.uk
http://www.livingtreeeducationalfoundation.org
http://www.livingwithclimate.fi
http://www.overstory.org
http://www.plant-trees.org
http://www.recoftc.org
http://www.ruaf.org/index.php
http://www.sactree.com
http://www.treesofstrength.org
http://www.unl.edu/nac/workingtrees
http://www.worldagroforestry.org

Index

A

Acacia albida, 43
Acacia arabica, 19, 20
Acacia auriculiformis, 19, 41, 44, 50, 53, 57, 73, 74, 75, 77, 79
Acacia catechu, 20, 39, 41, 50, 78, 97, 133
Acacia mangium, 41, 43, 44, 53, 57, 74, 88, 128, 129, 131
Acacia mearnsii, 39, 78
Acacia modesta, 30, 50, 78
Acacia nilotica, 39, 44, 53, 74, 75, 78, 79
Acacia senegal, 43, 78
Acacia tortilis, 43, 77, 78
Acer campbelii, 78
Acer oblungum, 78
Acrocarpus fraxinifolius, 78
Adansonia digitata, 43
Adina cordifolia, 28
Aegiceras carniculatum, 29
Aegle marmelos, 79
Aesculus indica, 78, 97
Afforestation, 7, 15, 153
Agrisilviaquaculture, 56
Agrisilvicultural system, 37, 54, 93
Agrisilvipastoral system, 37, 54
Agroforestry benefits, 10
Agroforestry diagnosis and design, 104–113
Agroforestry limitations, 11
Agroforestry need, 6–7
Agroforestry objectives, 5–6
Ailanthus excels, 19, 134
Albizia chinensis, 45, 78
Albizia lebbeck, 44, 50, 53, 78
Albizia procera, 19, 20, 78
Allelopathic interaction, 66
Alley cropping, 37, 41, 42, 54, 63, 67, 71, 73, 77, 131
Alnus nepalensis, 50, 66,
Alnus nitida, 78
Alpine forests, 31, 32
Amoora cuculata, 29
Anacardium occidentale, 39
Anogeissus latifolia, 97
Anthocephalus chinensis, 19
Apisilviculture, 56, 57
Aquaforestry, 56
Areca catechu, 78
Artocarpus heterophyllus, 74, 78
Avicennia alba, 29
Azadirachata indica, 19, 20
Azadirachta indica, 43, 49, 50, 53, 75, 77, 78, 92, 144

B

Backyard planting, 15
Bamboo, 2, 29, 35, 39, 87, 89, 102, 124, 160, 161, 163, 164
Bambusa balcooa, 160, 161
Bambusa bambos, 160, 161
Bauhinia variegate, 19
Bauhuinia purpurea, 19
Beach forest, 28
Bending, 98, 137
Bischofia javanica, 79
Bombax ceiba, 39, 78, 98, 134
Bombax malabaricum, 53
Boundary markings, 37, 51
Bruguiera conjugata, 29
Bruguiera parviflora, 29
Bushing, 98
Butea monosperma, 19, 20

C

CAFRI, 123
Calliandra calothyrsus, 43, 49, 73, 74, 75
Callophyllum littoralis, 29
CAMCORE, 120
Canal banks plantation, 16, 20, 21
Canopy density, 8, 22, 24
Cassia fistula, 19, 20, 51
Cassia javanica, 20
Cassia siamea, 19, 43, 44, 67
Casuarina equsetifolia, 43
CATIE, 120, 121
Cedrus deodara, 78
Ceiba pentandra, 39
Celtis australis, 78
Ceriops decandra, 29
Ceriops tagal, 29
Chloroxylon swietinia, 97
Chukrasia tubularis, 78
Classification of forest, 22
Cocus nucifera, 78
Community woodlots, 16, 18, 21
Competitive interaction, 61
Complementary interaction, 60
Coppicing, 22, 42, 54, 73, 77, 85, 97, 148, 153
Cordia alliodora, 45, 122
Cordia dichtotoma, 53
Corylus colurna, 78
Cryptomeria japonica, 39
CSIRO, 121
Cutting back, 49, 85, 97, 145, 156
Cynometra ramiflora, 29

D

Dalbergia latifolia, 28, 78, 79
Dalbergia sissoo, 19, 20, 39, 43, 44, 50, 57, 74, 78, 79, 131, 132
Delonix regia, 19, 20, 50
Dendrocalamus strictus, 79, 160, 161, 164
Depletion of forest, 7
DFSC, 121, 122
Dry alpine scrubs, 31, 32
Dry evergreen forests, 29, 30
Dry tropical forests, 29

E

Emblica officinalis, 53, 79
Environmental adaptation, 70
Erosion control, 7, 32, 33, 51
Erythrina berteroana, 53
Erythrina poeppigiana, 45, 75, 76
Establishment phase, 99
Eucalyptus camaldulensis, 78, 135
Eucalyptus citridora, 78
Eucalyptus globules, 78
Eucalyptus grandis, 78, 135, 137
Eucalyptus teriticornis, 78, 135
Excoecaria agallocha, 29
Extension forestry, 16, 17, 21

F

FAO, 117
Farm forestry, 13, 16, 21
Ficus bengalensis, 20
Ficus glomerata, 19, 79
Fixed-lift pruning, 97
Forest coverage, 8, 26
Forest types of India, 27–32
Fresh water swamp forests, 28, 29
Fuelwood crisis, 7

G

Glircidia sepium, 79
Gmelina arborea, 41, 43, 44, 50, 57, 77, 78
Grevillea robusta, 45, 49, 50
Grewia asiatica, 149,150
Grewia optiva, 78, 98, 149, 151
Grewia tilliaefolia, 28
Growing stock, 22, 24

H

Hardwickia binata, 79, 97, 98
Harvesting phase, 99, 100
Heritiera minor, 29
Himalayan dry temperate forest, 31
Himalayan moist temperate forest, 31
History of agroforestry research, 114
History of Indian forests, 24
Homegardens, 40, 54, 55, 56, 57, 72, 86, 100

I

IBPGR, 117
ICRAF, 2, 104, 105, 110, 115, 123
IDRC, 115, 121
ITTO, 117, 118
IUCN, 118
IUFRO, 119

J

Jacaranda ovalifolia, 20
Juglans regia, 78
Juniperus macropoda, 78

K

Kandelia candel, 29
Khaira, 133

L

Layering, 82, 85, 86, 138, 145, 147, 159
Leucaena leucocephale, 43
Linear strip plantations, 17
Littoral and swamp forests, 28
Lopping, 62, 73, 74, 76, 98, 133, 140
Lumnitzera racemosa, 29

M

Macro D & D, 105, 106, 108, 111, 113
Madhuca indica, 20, 79
Mangifera indica, 19, 20, 41, 79
Mangium, 41, 43, 44, 53, 57, 74, 88, 128, 129, 130, 131
Mangrove forests, 28, 29
Maturing phase, 99, 100
Method of regeneration, 22
Michelia doltsopa, 39
Micro D & D, 105, 108, 109, 111, 113
Microclimate improvement, 63, 64
Microclimatic modification, 59, 65
Mixed forestry, 17
Moderately dense forest, 8, 24, 27
Moist alpine scrubs, 31, 32
Moist tropical forests, 28
Montane wet temperate forests, 31
Moringa oleifera, 77, 78
Morus alba, 56, 57, 78, 98, 151, 152
Morus indica, 78
Morus nigra, 151
Morus rubra, 151
Mulberry, 57, 151, 152
Multipurpose trees, 37, 43, 50, 51, 55, 68, 71, 72, 74, 109, 128–164

N

National agroforestry policy, 124, 125, 127
Neem, 144, 145, 146
Negative interactions, 62, 63, 65, 66, 67
Nitrogen fixing trees, 49, 63, 76, 143
Nursery management, 86
Nutrient cycling, 4, 6, 62, 63, 77, 124

O

Oak, 31, 158, 159, 160
OFI, 122
Olea cuspidata, 30
Open forest, 8, 24, 26, 27,

P

Parkia biglobosa, 43
Phalsa, 149, 150
Phoenix dactifera, 41
Phoenix paludosa, 29
Pine forests, 30, 31
Pinus alata, 78
Pinus gerardiana, 78
Pinus khasya, 31
Pinus roxburghii, 31, 50, 78,
Pinus wallichiana, 78
Pollarding, 67, 73, 85, 97, 98, 148, 153
Polyalthia longifolia, 19
Pongamia pinnata, 19, 78
Poplar, 153, 154, 155, 156, 157, 158
Populus alba, 78
Populus ciliata, 78, 153
Populus deltoides, 78, 153, 154, 155, 158
Populus euphretica, 78
Populus nigra, 78
Populus tremula, 78
Positive interactions, 62, 63, 64, 65
Preplanting phase, 99
Pricking out, 87, 88, 89, 129, 144
Prosopis chilensis, 78

Index 171

Prosopis cineraria, 63, 74, 77, 78
Prosopis juliflora, 19, 20, 44, 50, 75
Protein bank, 52, 53
Pruning, 32, 41, 42, 47, 49, 63, 64, 67, 71, 72, 73, 74, 76, 81, 85, 87, 95, 96, 97, 100, 101, 129, 130, 139, 147, 148, 151, 157, 164
Pterocarpus dalbergioides, 39
Pterocarpus marsupium, 79
Pterocarpus santalinus, 78

R

Railway line plantation, 19, 20
Ravine lands, 14, 15
Recreation forestry, 16, 18, 21
Rhizophora mucronata, 29
Roadside plantation, 18, 19
Robinia pseudacacia, 78
Rotational woodlots, 37, 50

S

Salix alba, 78
Salix fragilis, 78
Salix tetrasperma, 29
Salvadora persica, 78
Samanea saman, 19
Santalum album, 39
Saraca indica, 19, 20
Scarification, 84, 144
Schima wallichii, 39, 78
Selective pruning, 97
Self-pruning, 71, 81
Sesbania grandiflora, 53, 55, 74, 76, 77
Sesbania sesban, 43, 49
Shade trees, 45, 64, 69, 92, 93
Shelterbelts, 17, 33, 37, 45, 46, 47, 64, 69, 75, 77, 137
Shifting cultivation, 25, 27, 37, 38, 39, 40, 115
Shorea robusta, 28, 30, 39
Silvipastoral system, 51, 52, 53, 54
Sissoo, 19, 20, 39, 43, 44, 50, 57, 74, 78, 131, 132
Social forestry, 13–21
Social forestry characteristics, 13
Social forestry components, 16–18
Social forestry objectives, 14

Social forestry scope, 14–16
Soil conservation, 15, 17, 37, 50, 54, 55, 56, 57, 62, 64, 68
Soil conservation hedges, 37, 50
Sonneratia acida, 29
Sonneratia caseolaris, 29
Stand density, 22, 24
Status of Indian forests, 22–35
Stratification, 36, 84
Sub-alpine forests, 31
Subtropical montane forest, 30
Supplementary interaction, 60
Syzygium aromaticum, 41
Syzygium cumini, 19, 79

T

Tamarind, 19, 39, 43, 53, 77, 79, 147, 148
Tamarindus indica, 19, 39, 43, 53, 77, 79, 147
Taungya system, 37, 38, 39, 40, 50
Tecomella undulate, 77
Tectona grandis, 28, 30, 39, 43, 76, 78, 79, 140
Temperate montane forest, 31
Terminalia alata, 78
Terminalia arjuna, 19, 20
Terminalia catappa, 19
Terminalia myriocarpa, 78
Terminalia paniculata, 28
Terminalia tomentosa, 28
Thinning, 81, 93, 94, 95, 96, 100, 129, 132, 133, 134, 136, 139, 142, 163, 164
Tip pruning, 97
Toona ciliata, 78
Tree management, 67, 80, 91, 129, 132, 133, 136, 138, 142, 144, 147, 149, 152, 156, 159
Tree-animal interface, 64, 66
Tropical dry deciduous forest, 29, 30
Tropical dry evergreen forest, 29
Tropical moist deciduous forest, 28
Tropical moist evergreen forest, 28, 29
Tropical moist semi-evergreen forest, 28
Tropical thorn scrubs, 30
UNEP, 119, 120

V

Variable-lift pruning, 97,
Vateria indica, 20
Vitellaria paradoxa, 43

W

WAC, 115, 116
Wastelands, 6, 8, 14, 15, 17, 25, 123
Wet hill broad leaved forest, 30
Wildings, 83, 84, 87, 89

Windbreaks, 33, 37, 46, 47, 48, 49, 50, 53, 57, 64, 69, 75, 93, 137, 149
Woody hedgerows, 54, 64

X

Xylocarpus granatuns, 29
Xylocarpus molluccensis, 29